KU-727-321

The Hamlyn All-Colour Animal Encyclopedia

Published 1981 by
The Hamlyn Publishing Group Limited
London · New York · Sydney · Toronto
Astronaut House, Feltham, Middlesex, England.
© Copyright The Hamlyn Publishing Group Limited 1981

All rights reserved. No part of this publication
may be reproduced, stored in a retrieval system,
or transmitted in any form or by any means,
electronic, mechanical, photocopying, recording or
otherwise, without the permission of The Hamlyn
Publishing Group Limited.

ISBN 0 600 30370 5   Printed in Italy by Interlitho

The Hamlyn All-Colour Animal Encyclopedia

Cathy Kilpatrick and John Hard

HAMLYN London·New York·Sydney·Toronto

ACKNOWLEDGEMENTS

Transparencies

Heather Angel, Farnham 36, 52, 65 top, 70, 71, 72 bottom, 88, 90 top, 91, 93, 98, 111, 117, 119 bottom, 120, 122-123, 125 top, 125 bottom, 128, 131, 135, 142, 148, 162 bottom, 184, 230 right, 235, 315 bottom;

Aquila Photographics, Studley 154, 191 top, G. V. Hawksley, 170 bottom, Gary Weber, 157, 197 bottom;

Ardea Photographics, London 158 bottom, I. R. Beames front jacket, 49, 56 top, Graeme Chapman 208 top and bottom, K. W. Fink front jacket, inset bottom left, 170 top, 201, Clem Haagner 250 top, P. Morris front jacket, inset top far left, 9 top, 62 top, 107, 223, 232, C. K. Mylne 209 top, Ron and Valerie Taylor 26, 103, Adrian Warren 252;

Biofotos—S. Summerhays 63;

S. C. Bisserot, Bournemouth title page, inset centre, 72 top, 73 top, 79, 81, 82 top, 89, 115, 124, 144 bottom, 145, 150 top, 156-157, 227, 230 left, 237;

Bruce Coleman, Uxbridge 138 bottom, 208, 243, 257, 289 bottom, 293 bottom, 305 bottom, Helmut Albrecht 244, Ken Balcomb 246, Jen and Des Bartlett 31 top, 137 bottom, 236, 242 top, 284 top, S. C. Bisserot 39, 122, R. and M. Borland 66, Jeffrey Boswall 200, Mark Boulton 22, 303, Jane Burton 20 bottom, 42 top, 42 bottom, 75 bottom, 87 top, 94, 114 bottom, 120-121, 228, 254 top, 267 top, 275, 280 top, 283, 310 top, Bob and Clara Calhoun 35 top, 266, 312 bottom, R. I. M. Campbell 242 bottom, 307, Alain Compost 295 bottom, Eric Crichton 312 top, Gerald Cubitt 33, 311, A. J. Deane 57, Jack Dermid 193 bottom, Francisco Erize 14 left, 14 right, 41, 222-223, 224, 247 bottom, 248 top, 284 bottom, J. Fennell 9 bottom, Jeff Foott 261, 264, Dian Fossey 245 centre, M. Freeman 254 bottom, C. B. Frith 234, Francisco Futil 182-183, Mary Grant 136, Jerry L. Hunt 306, Jon Kenfield 116, Stephen J. Krasemann 249, Gordon Langsbury 251, Norman Lightfoot 31 bottom, 271 top, Lee Lyon front jacket, inset top right, 7 top, 245 top, 245 bottom, 292 top, John Markham 241, Norman Myers 212-213, 255, 263 bottom, 268 bottom, 309 bottom, M. Timothy O'Keefe 104, 276, Oxford Scientific Films 253 bottom, Roger Tory Peterson 13, Graham Pizzey 17, 220 bottom, 225 bottom, G. D. Plage 285, 291, Mike Price 286 top, Hans Reinhard 11, 12, 48, 80, 106, 226, 253 centre left, 262 top, 290, 297 top, 297 bottom, 304, Leonard Lee Rue III 218, 256-257, 259 bottom, 270 bottom, 281 bottom, W. E. Ruth 298, Frieder Sauer 60, Horst Schaffer 288, John Shaw 21 top, 75 top, Stouffer Productions 45 bottom, 253 centre, 280 bottom, Barrie Thomas 268, Norman Tomalin 15, 40, 238, 239 bottom, 265 top, 265 bottom, D. and K. Urry 45 top, Joe van Wormer title page, 10, 47, 172, 262 bottom, 308, Rod Williams 267 bottom, Bill Wood 7 bottom, 87 bottom, Jonathan Wright 6 bottom left, Gunter Ziesler 78, 110, 176 bottom, 233, 248 bottom, 295 top, Christian Zuber 192;

Adrian Davies, Wallington 174 bottom, 181 left;

Robert Gillmor 181 right;

Brian Hawkes, Sittingbourne 21 bottom, 30, 183, 206-207, 209 bottom, 274;

D. P. Healey, Middlesborough 126, 134;

David Hosking, London 18, 51 top, 112 bottom, 179, 259 top, 282;

Eric Hosking, London back jacket, 6 right, 22-23, 35 bottom, 37, 55, 140 bottom, 146, 150 bottom, 152-153, 161, 163 bottom, 164, 165, 168, 174 top, 175, 176 top, 185 bottom, 191 bottom, 196, 204, 206, 211 top, 211 bottom, 229, 240, 250 bottom, 260, 272, 287, 300;

Jacana, Paris 155, 158 top, 278 bottom, Devez Cnrs 133 top, Andre Ducot 195, Brian Hawkes 178, P. Laboute 151 bottom, Varin Visage 133 bottom;

Frank Lane, Pinner 199, Peter David 68;

Mansell Collection, London 182, 277;

Natural History Photoraphic Agency, Saltwood 210, Douglas Baglin 141 top, 144 top, 216, 221 bottom, 222, 225 top, Anthony Bannister 34, 69, 73 bottom, 83 top, 84, 129 bottom, 138 top, 143 bottom, B. Barnetson 119 top, Joe B. Blossom 185 top, N. A. Callow 82 bottom, 83 bottom, Stephen Dalton 6 top left, 58, 65 bottom, 74, 77, 127, 143 top, 188-189, 203, 207, Robert J. Erwin 151 top, Brian Hawkes 162 top, 256, 299, 314, 315 top, E. A. Janes 20 top, Peter Johnson endpaper, Roy D. Mackay 217, M. Morcombe 147 top, 205, W. J. C. Murray 278 bottom, K. B. Newman 149 top, L. H. Newman 220 top, Roger Perry 16, E. Hanumantha Rao 139, 239 top, 263 top, 273, Philippa Scott 140 top, James Tallon 187, 302, M. Tomkinson 293 top, A. J. Van Aarde 270 top, P. Wayne 247 top;

Natural Science Photos, Watford—M. E. Bacchus 129 top, C. Banks 149 bottom, 221 top, Isobel Bennett and P. G. Myers 62 bottom, P. Boston 193 top, P. J. K. Burton 186, H. Dossenbach 202, Nat Fain 27, A. Grandison title page, inset left, J. A. Grant 76, F. Greenaway 178-179, J. M. Hobday 173, Ricard Kemp 289 top, 309 top, G. Kinns 292 bottom, D. B. Lewis 95, 163 top, Gil Montalverne 100, Stephen Morley 305 top, L. E. Perkins 118, M. R. Stanley Price 147 bottom, 310 bottom, D. M. Turner-Ettlinger 160, 180, C. A. Walker 141 bottom, 166, 167, 197 top, P. H. Ward 85, 132, 137 top, 156, 281 top, 296, Curtis E. Williams 194;

Oxford Scientific Films, Long Hanborough 50, 86 top, 86 bottom;

Photo Aquatics, Theydon Bois—Carl Roessler 271 bottom;

R.I.D.A. Photographic Library—Miles Harrison front jacket, inset top left;

Seaphot, Bristol—Roberto Bunge 24, Peter Capen 109, Richard Chesher 90 bottom, Dick Clarke 64, 97, Peter David 112 top, Walter Deas front jacket, inset bottom right, 28, 51 bottom, 102, Christian Petron 101, Rod Salm 114 top, Warren Williams 67;

J. Wagstaff title page, inset right;

Worldwide Butterflies—Robert Gooden 56 bottom.

Illustrations

Ray Burrows; Creative Cartography Limited; Linden Artists Limited; Susan Neale; The Tudor Art Agency Limited; Mike Woodhatch (David Lewis Artists).

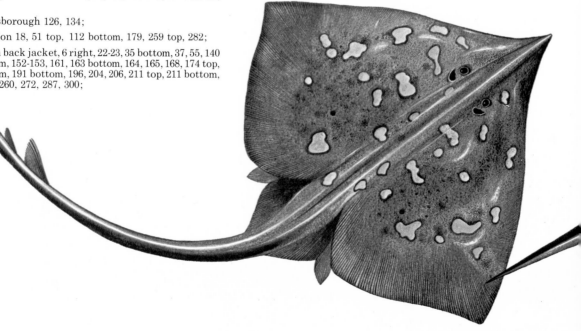

4

Contents

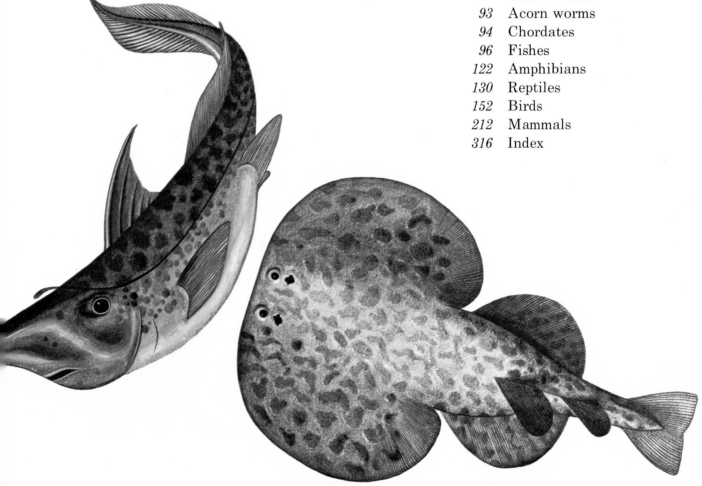

Introduction

There are over a million species of animal on our planet. They range in size from the minute blood parasites that invade man and are only a few hundredths of a millimetre across, to the giant whales which can grow to nearly 30 metres in length.

The range of habitats in which the different animals are found is

just as great as the extremes in size. All the oceans contain an abundance of life, from the drifting jellyfishes and larval forms that live close to the surface to the strange worms and fishes that dwell in the black depths of the abyss, a region with a constant temperature that is only just above freezing, and with no light at any time.

The rivers and lakes have their own animal life, or fauna, with numerous insect and crustacean forms amongst the stones and weeds, fishes throughout their depths and many birds and rodents along their banks.

The land masses have a greater variety of habitats for animals than have the waters as their temperatures vary more and there is the ever changeable wind and rain. The mountains hold plenty of water, but it is trapped in ice and snow and little of it is available to animals and plants; most of their water requirements come from the clouds in the form of rain. A mountain habitat also poses problems associated with lack of oxygen and the biting, cold, winds. Only well adapted mammals, such as the yak, the alpine ibex and the llama, and a few birds, like the condor, lammergeyer and alpine chough, are found in the high peaks. Other animals would become dormant in the extreme cold.

Top: *Army ants moving in their usual formation when on the march, with savage mandibles ready to seize any unfortunate prey. Their protection lies in the fact that despite being small animals they work together in vast hordes, the death of an individual being unimportant.*
Above: *These superb mustangs are well adapted to life on the open prairies of North America or the vast plains of Central Asia. In this picture two young males are showing the fighting value of quick reflexes and flying hooves.*
Right: *The great horned owl (Bubo virginianus) of North America. The large eyes and superb hearing enable owls to catch such prey as mice and rats which feed during the hours of darkness. The characteristic tufts of feathers of the horned owls are not ears but help to break up the outline.*

The jungles swarm with life, since the temperature and humidity remain high throughout the year and even the most primitive invertebrates can live without the problems that arise with temperature fluctuations. Unfortunately man has cleared some of the finest forest and jungle regions and this has caused many species to dwindle in numbers and even to become extinct. For example, the island of Madagascar has lost over 80 per cent of its native forest and its unique fauna is fast disappearing.

The open forests and grasslands each have their own types of animal and these vary in the different parts of the world, although they exhibit similar ways of life. In Africa the grass-eating animals are well known, gazelles, zebra, wildebeest for example. In South America there are fewer deer but many more rodents; while in Australia the kangaroo and wallaby are the grazing species. The actual species found on a particular land mass depends upon the evolutionary history of that continent.

One of the more difficult habitats for animals to cope with is the desert. Despite the problems of water shortage and the tremendous daily temperature fluctuations, many small insects, spiders and scorpions are to be found in the deserts, living off the scanty dry vegetation, seeds and, of course, each other. These small arthropods form an important part of the diet of the desert reptiles, rodents and birds.

Probably the worst, and as yet uninhabited, site is the dry desert in central Antarctica where no ice or snow is found. There are only dry grains of sand covered with a film of salt. The temperature here is below freezing, there is no moisture and no life.

Later in this book there are descriptions of the different habitats and some of the animals that live in them. Then there follow accounts of the various groups into which the incredible number of animals has been classified.

Above: *A herd of African elephants showing a number of females, with 2 youngsters, which have spent some time drinking and rolling in the cooling mud. The liquid mud not only cools these great mammals but helps in the removal of some parasites and insects from the thick hide.*

African elephants spend much of their life wandering the plains in search of food. In some game reserves their numbers have risen enormously causing vast areas of land to become semi-desert following overgrazing and subsequent land erosion. Although it is necessary to keep down the numbers, poaching results in the lingering death of hundreds of elephants.

Right: *One of the many attractive feather stars waving its numerous graceful arms as it moves over the surface of a coral reef. The coral reefs provide some of the most colourful animals and at the same time are the site of much savage competition and death (see page 87).*

Feather stars are found throughout the warmer oceans of the world, some in shallows whilst others are to be found in the dark abyssal depths.

Major biogeographical regions

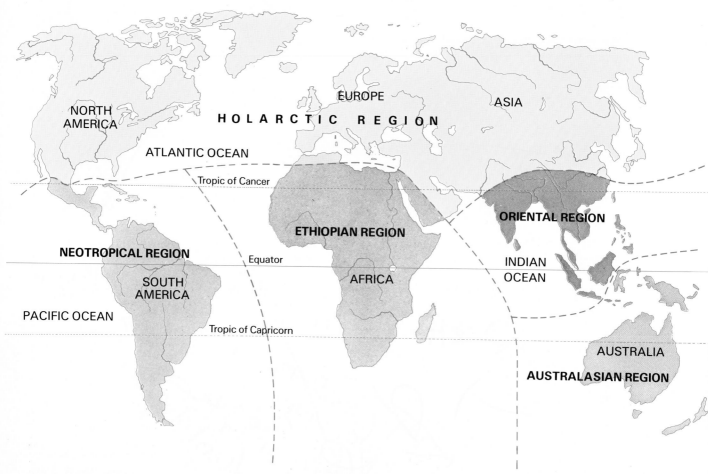

The world is divided into a number of major regions, or realms, each having its own special animal types. The kinds of animals found in any region depend on two main features – the climate and the plant life. Naturally the climate is determined to some extent by the country's proximity to the Polar regions or the equator, and correspondingly the plant life, for which rainfall and temperature are very important, is controlled by the climate.

The world is divided into 5 main biogeographical regions: the **Holarctic** region – made up of North America, Europe, part of North Africa and much of Asia; the **Ethiopian** region – most of Africa and part of Arabia; the **Oriental** region – India, Malaysia, Burma, South China; the **Neotropical** region – Central and Southern America; and the **Australasian** region – Australia, New Guinea, New Zealand and some of the South Indonesian islands.

The largest barriers between these different regions are oceans and mountains, although the great Sahara Desert separates the Ethiopian region from the southern limits of the Holarctic region. The barriers we see today may not always have existed, and many widely separated areas were once closer together. Many scientists believe that they split apart more than 100 million years ago and 'drifted' away from each other. Thus South America and Africa were once linked, but South America has moved westwards creating the south Atlantic Ocean. About the same time Antarctica broke away and moved

Right: *Plants such as these fossilized ferns thrived some 300 million years ago in dense humid forests. Their remains have provided one of the most important fossil fuels, coal. The veins of these fossil leaves can still be seen.*

south to its present position.

The evidence for these continents once having been close together is found in the layers of ancient rocks. The Antarctic has layers of coal, which are the fossilized remains of extinct plants that once lived in a warm climate. Similar coal deposits are to be found in the continents of Africa and South America. If the continents have drifted apart, it means that they may have had some ancient form of animals in common. Newer forms, however, would probably be different, unless they were able to migrate vast distances. The longer an area is separated from all other land the greater the divergence in its animal life. The best example of this is Australia.

Right: *Ammonite fossil. The preserved remains of extinct animals such as ammonites have helped man to trace the changes in evolving animals. The ammonites are extinct relatives of the squids.*

Holarctic region

This is easily the largest biogeographical region and is divided into two parts – North America (Nearctic) and Eurasia (Palearctic).

The different types of country to be found in the two regions of the Holarctic are very varied. In the north are the frozen wastes of the Arctic and cold tundra. Then come the dark coniferous forests of Canada, North America and northern Europe. There are also forests of deciduous trees rich in various forms of animal and plant life. Next there are the open grasslands of the Russian steppes and American prairies. The lakes and rivers are numerous, and afford a home for fishes, frogs, aquatic insects and crustaceans, while the deltas and swamplands of Europe and the southern United States teem with bird and fish life.

The mountains and general terrain of Eurasia are far more varied than those of North America and hence there are many more different habitats in which animals can live. For this reason there is a greater number of species in the eastern Palearctic than in North America. Thus, for example, there are many species of goat and sheep in Eurasia, whereas there is only a single species of each occurring naturally in North America. This is also true of oxen and deer, there being far more different species in Europe and Asia than in North America.

There are close resemblances between the animals of North America and the rest of the Holarctic, and this is due to the fact that there was once a land link joining the northern region across what is now the Bering Straits, and over which the animals could migrate. Once established in their new homes the different species flourished until the ice

KEY

Alpine tundra and ice-desert

Coniferous forest

Broad-leaved forest and meadow

Evergreen trees and shrubs

Temperate rain forest

Tropical rain forest

Thorn forest

Grassland

Scrub, steppe and semi-desert

Desert

Above: *Raccoons in the wild spend a lot of their time near water, hunting food. They can climb trees and hibernate in hollows of the trunk. They have also adapted to scavenging on man's rubbish.*

and snow of the Ice Age forced them south. In America some of the northern species moved south down through Central America and into South America and then returned when the ice cap melted. In Eurasia, the return migrations were not so clear.

Some animals evolved in one half of the Holarctic, migrated to the other half and stayed there, while becoming extinct in their place of origin. This is seen very clearly in the past history of the camel and horse families. These families evolved in North America, moved into Eurasia and flourished in the different habitats, meanwhile dying out in North America. The horse was reintroduced into North America, of course, with great success. The raccoon family developed in North America, spread into Eurasia and then retreated before the ice cap. In North America, the raccoons recolonized the land following the thawing of the snow and ice, whereas

10

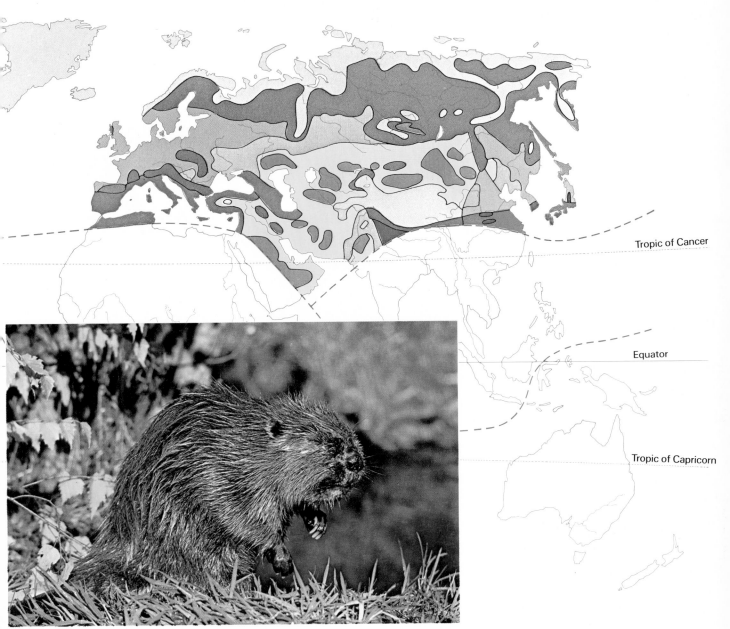

Above: *Because of their dams, beavers are very useful in the natural control of waterways throughout the Holarctic forests.*

in the Palearctic there are no living forms. The giant panda (*Ailuropoda melanoleuca*) of the northern Oriental region is their closest relative. However, the similarities between the animals of the Nearctic and Palearctic are very marked. Neither contain any marsupials, and monkeys are absent in the west and only just present in the southern and eastern borders of the Palearctic.

The carnivores are very similar, wolves, foxes, lynxes, weasels and polar bears being found in both. Similarly rodents such as rats, mice, voles, lemmings and beavers are found throughout the region. The great bison of Europe is closely related to the North American species.

There are similarities in the fishes and amphibians found in the numerous rivers and lakes in both the great land masses. These waters are relics of the melting snows, and are replenished by the mountain drainage.

Birds, with their superb power of flight, are found right across the Holarctic with closely related or even identical species occurring over thousands of kilometres. Gulls and seabirds are found along the North Atlantic seaboard of both sides of the Atlantic, and wildfowl are present along the rivers and lakes of Canada and America as well as the cold lakes of northern Europe and the rivers of Asia. The birds of prey are also closely related — the eagles of the mountains, the owls of the forests and fields, and the hawks of the moors. A large number of species are found throughout the Holarctic.

11

Ethiopian region

Bordered in the north by the immense dry waste of the Sahara Desert, the Ethiopian region has some of the most splendid and best-known of all animals – animals such as the lion, elephant, gorilla, giraffe and ostrich. Africa has an extremely varied range of terrain and climate, but there are only a few major geographical features that influence the habitats.

Firstly, there is the enormous Sahara Desert and the large area to the south of the Sahara that produces only poor scrubland and sparse grass. This desert and semi-desert supports few large animals, the animal life being mainly restricted to insects, scorpions, lizards, snakes, desert rats, foxes and a few birds. Similar animals are found in and around the terrible Namib Desert of south-west Africa.

Dividing Africa like a wide belt is the tropical jungle or rainforest. The plants are luxuriant and grow up towards the light, forming a dense canopy, and it is in the canopy that most of the animals live. This area teems with monkeys, squirrels and various rodents, all busily eating the leaves, seeds and fruit. There are few grazing herbivores since there is very little grass or fresh greenery on the ground. The leaf litter, so common in deciduous woodlands of Europe, is missing, for it is broken down incredibly quickly by a thriving population of fungi and insects. Decaying timber is quickly taken by the termites which are everywhere. The most obvious and noisy members of the jungle's fauna are the birds – parrots,

sunbirds, hornbills, toucans – all of which are splendidly coloured. Birds of prey, such as sparrowhawks, owls and goshawks, all capable of manoeuvring around the trees, feed on the other birds and rodents.

Two other important regions of Africa are the scrubby bushveldt and the open grassland or savanna. These two areas blend into one another and hence some animals, for instance the elephant, cheetah, giraffe and lion are common in both. The bush is characterized by extremely dry conditions and the plants are specialized to withstand drought by being tough and leathery, with the leaves thin and thorny. The two periods of rainfall, in August and November, bring remarkable changes in the appearance of the plants, which blossom and produce fruits and seeds in a very short time.

The change in the grassy savanna is similarly dramatic when rain comes, but trees and bushes are less common since there is only a relatively thin layer of soil, with hardened volcanic lava underneath. This prevents the development of plants with deep roots. The open savanna provides perfect conditions for grazing herbivores, and here there are the largest herds of antelope in the world. Similarly vast herds of zebra, gazelle and gnu move slowly over the whole region. The numerous herbivores do not usually compete for food, however, as they eat different levels of vegetation. Zebras eat long grass, gnus eat slightly shorter grass, gaz-

Right: *Lions are social animals that live in groups or prides. Several females and their cubs live together with the dominant male. Here the pride are resting in the shade from the heat of the day.*

elles feed on very short grass, whilst the hartebeests graze on tough stalks and tufts left by the others. A similar method is seen in the leaf feeders – the giraffe eats very high leaves and twigs, the gerenuk stands on its hind legs and nibbles low branches and leaves, whilst the shorter dik-dik eats leaves only a few centimetres above the ground.

This abundance of herbivores means that there are a great many carnivores which catch and kill them for food, and following these hunters are the scavengers, such as vultures and hyenas, which eat up the remains.

The great rivers teem with animals, ranging from the huge hippopotamus and crocodile to the elegant waterfowl, such as the heron and flamingo.

Below: *Flamingoes are colonial birds and build their cone-shaped nests from soft mud which hardens.*

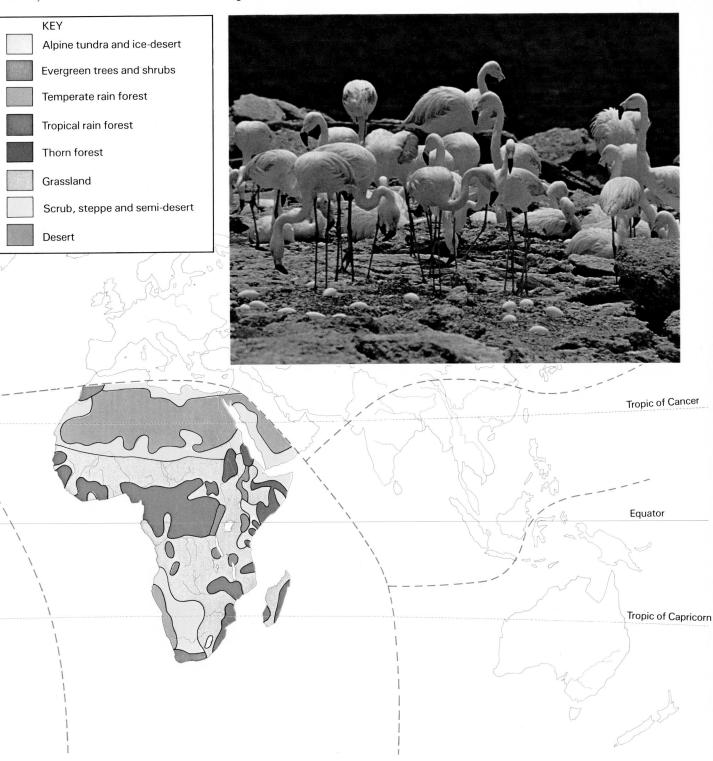

KEY

- Alpine tundra and ice-desert
- Evergreen trees and shrubs
- Temperate rain forest
- Tropical rain forest
- Thorn forest
- Grassland
- Scrub, steppe and semi-desert
- Desert

Tropic of Cancer

Equator

Tropic of Capricorn

Neotropical region

The Neotropical region consists of 3 main habitats: the tropical rainforest (Amazonian); the Andes mountain chain, the largest in the world; and the temperate grassland (pampas). There is also a variety of other habitats – steppe with regular seasonal rainfall, high plateau country, dry desert and southern tundra.

The mixture of wildlife (marsupials are found together with placental, or higher, mammals such as ocelot and deer) is due to its history. Over 70 million years ago a land bridge existed, linking the region with North America and animals crossed over it into South America. Later the land bridge was submerged by water, and South America became an enormous island. The existing species of that time flourished and diversified. Three to four million years ago the Panama land bridge was re-established and there was a second invasion of animals. These new invaders were often stronger than the existing residents, many of which died out since they could not compete successfully.

No other continent has such a large proportion covered by dense forest. It has very high humidity with an enormous number of tree species and many other smaller plants growing on the larger ones. There is dense shade and black soil, rich in litter. Due to regular flooding, most of the animals can either swim (fishes, reptiles, rodents and tapirs) or climb (spider monkeys, porcupines and sloths). Some, like the

Life at the top

The mountainous region – the Andes – is so vast that it is difficult to describe a typical area. It consists mainly of very high plains or plateaux, which are split by deep valleys, and surrounded by rocky crags and slopes, whilst in the background loom the snow-covered peaks. Two of the great problems are the shortage of oxygen at the high altitudes, much of the area being between 3000 and 4000 metres, and the great variation in temperature. The llama (*Lama glama*) and vicuna (*Lama vicugna*) have successfully adapted to these conditions.

Below: *Capybaras* (Hydrochoerus hydrochaeris) *are the largest rodents in the world and are found on the pampas especially near water or marshes.*

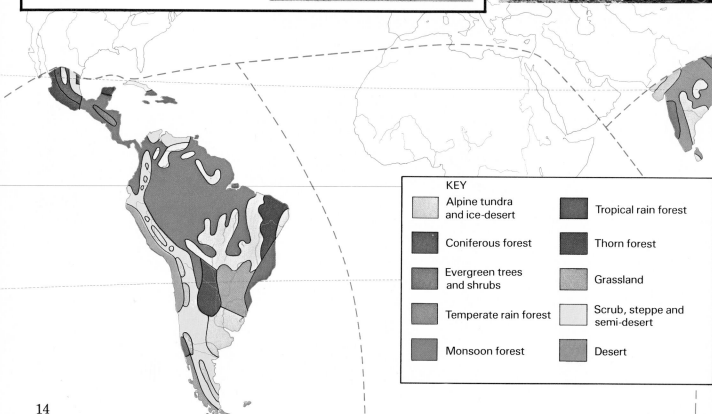

KEY

Alpine tundra and ice-desert	Tropical rain forest
Coniferous forest	Thorn forest
Evergreen trees and shrubs	Grassland
Temperate rain forest	Scrub, steppe and semi-desert
Monsoon forest	Desert

jaguar, both swim and climb.

The pampas consist of open grassland and areas covered only by scrubby bushes, with occasional trees, and occur mainly in the southern half of South America. This is an area having long, dry seasons followed by short, wet periods. In the extreme south the rainfall becomes more irregular; dusty dry deserts appear, with only patches of tough grass in hollows or sheltered regions.

The pampas are ideal grazing regions for herbivores, and include such species as the pampas deer. The typical South American herbivores of the pampas are rodents such as the plains viscacha, the Brazilian cavy and the capybara.

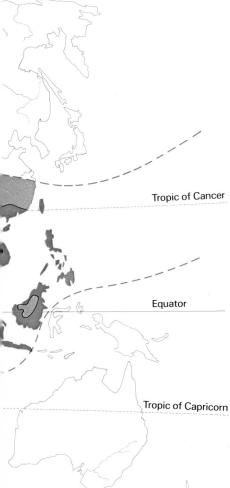

Tropic of Cancer

Equator

Tropic of Capricorn

Oriental region

The Oriental region has a wide variety of habitats, and consists of three main types of country: the hot, wet rainforests, the thorn scrub and savanna, and the mountain slopes. The mountains and deserts of the north border the Holarctic, whilst in the south-east the Oriental region exists as a mass of islands including the Malay archipelago and Indonesia.

Plant life, and therefore the animals that depend on the vegetation, is governed by the wind and rain – the monsoon. There are two monsoons – one blowing from the south-west in summer and the other blowing from the north-east in winter.

There is true jungle-like rainforest, for example in Assam and Borneo, where rainfall is heavy all year round and the temperature is always high. In this *primary* forest there are many fruit-eating and insect-eating animals as there is little light under the canopy and hence only scant grass or other plant life for animals to feed on. Typical animals are the gibbon, the orang utan, a few deer, numerous birds, bats and insects. At the coast, jungle is

replaced by mangrove swamp with its own unique fauna. In this lighter, *secondary* forest there is illumination through the canopy with some grass and bush growth and fewer trees. The gibbons and orang utans are replaced by langurs and macaques. This secondary forest is often a form of monsoon forest where there is heavy rain for 6 months of the year, followed by 6 months of drought.

In areas where rainfall is even further restricted, deciduous forest develops, so that the region looks very like parts of the Holarctic. If the rainfall is below 75 centimetres per year then the forests may give way to flat areas of scrub and thorn – similar to the bushveldt of Africa. Over large areas of Indo-China the land is covered in almost savanna-like grassland with clumps of thorn. The only dense vegetation is along the river banks. This open country is populated by grazing herds of herbivores like those found in the Ethiopian region, namely elephants, gaur and deer, and by carnivores such as wild dogs and leopards.

Below: *Gibbons are large apes with the ability to swing gracefully through trees using mainly their enormously long arms. Their arms are so long that walking on the ground appears awkward.*

Australasia

This region contains one of the most unusual collections of animals in the world, indeed many of the animals of Australia itself are unique. The fauna of the island chain in the north shows links with the fauna of the Oriental region, for instance tarsiers and jungle fowl are present, but the islands *immediately* to the north of Australia have more typically Australian animals such as the cockatoo, kangaroo and cassowary.

It is thought that the land links with the Oriental region separated many millions of years ago and the animals living on the Australian continent flourished and adapted. These were the pouched mammals, or marsupials, such as the kangaroo, wallaby and koala. It was not until man arrived only a few thousand years ago that some changes occurred. The important and damaging change has occurred in the last 200 years where man has altered the countryside by introducing livestock such as sheep and cattle, together with dogs, cats and rats.

The centre of Australia is desert and semi-desert, surrounded by arid scrub and patchy grassland. This makes up a third of the entire Australian mainland. There is an extensive savanna with scrub reminiscent of the African savanna. In the northeast and south-east there are belts of dense forest and on the mountain slopes there are woods and

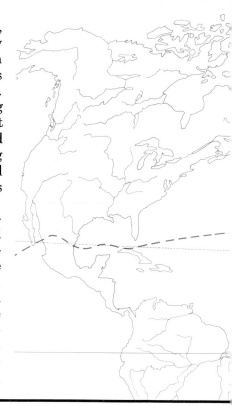

Below: *The Galapagos Islands are the site of many fascinating examples of evolution, including the finches which have evolved into 14 different species from one original form. The seed-eaters have thick, short bills and live on the* ground *a great deal and the insectivores have slender pointed beaks and live in trees. The woodpecker finch* (Camarhynchus pallidus) *uses a cactus spine to probe in holes to extract grubs — a counterpart to the beak of woodpeckers.*

Isolated islands

There are two main types of island - continental and oceanic. Continental islands are adjacent to large continental land masses, and their animals and plants are usually continental forms that have settled on the islands before, or after, their separation from the mainland. Oceanic islands are the result of volcanic activity out in the ocean and these islands are often isolated from all other land masses. Examples are Hawaii in the Pacific, and Tristan da Cunha and Gough Island in the south Atlantic.

Oceanic islands are seldom very large and their climate is governed by the ocean in which they are found. The range of habitats is limited usually to shoreline, rocks, sand and scrub, and only a few have any dense forest. The colonization of these islands relies on water currents and winds to bring seeds, fruits, spores, and drifting species to their shores. For example, Ascension Island receives species from Africa carried by the Benguela current. Because they can fly, birds, bats and insects are often the only animals present.

Fresh water is often very scarce, and hence those species of animal that require it, such as freshwater fishes and amphibians, are usually rare. Once an animal has established itself, it often evolves into a number of different forms in order to live successfully in the different habitats available. This is seen very clearly in the different finches of the Galapagos Islands.

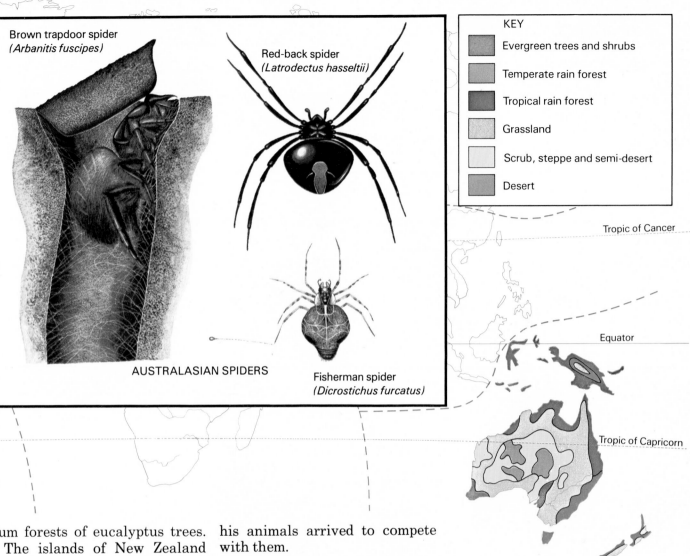

Brown trapdoor spider
(*Arbanitis fuscipes*)

Red-back spider
(*Latrodectus hasseltii*)

AUSTRALASIAN SPIDERS

Fisherman spider
(*Dicrostichus furcatus*)

KEY

Evergreen trees and shrubs

Temperate rain forest

Tropical rain forest

Grassland

Scrub, steppe and semi-desert

Desert

Tropic of Cancer

Equator

Tropic of Capricorn

um forests of eucalyptus trees. The islands of New Zealand ave their own climate and mixed ora, ranging from the mosses nd lichens of the high alps to the ense bush of the valleys. Much f New Zealand is covered in rassland.

The unique fauna of Australia ncludes the famous duck-billed latypus (*Ornithorhynchus ana- inus*) which lives in and along he rivers of east Australia. The 5 pecies of echidna, or spiny ant- ater, are other curiosities and, ike the platypus, are mammals hat lay eggs. These animals form he group known as the mono- reme mammals. The other un- sual group of mammals is that omprising the marsupials. These ormed a large group of animals, lling all the various habitats of ustralia before modern man and

his animals arrived to compete with them.

Below: *As herbivores, kangaroos play an important part in Australian ecology.*

Forests

Tropical forests

The tropical forests are found in a belt around the world, extending from the Amazonian rainforest, across the narrow band of the Congo and East Africa to the lush jungles of the Indo–Malaysian region. For such rich plant life it is necessary to have a high rainfall spread throughout the year, accompanied by high temperatures. This permits growth and reproduction at all times of the year, and hence provides for animals a variety of foods throughout the year. This means that animals can become highly specialized in their feeding without the danger of the food supply disappearing as the seasons change. Hence the nectar feeders, such as hummingbirds and hawkmoths, and the fruit-eaters, such as toucans and fruit bats, all flourish.

A feature of the tropical forests is the many different types of tree that give rise to the various levels of the canopy. At each level there will be different species of animal — insects and rodents in the bottom herbal layer; birds, insects, lizards and snakes among the saplings and bushes; and an enormous range of species in the upper tree layer.

There is very little grass in the jungles and most of the leaf litter is broken down so quickly that the ground appears bare when compared to the grass, herbs and drifts of leaves found in deciduous forests, or the damp carpet of pine needles found in the coniferous forests of the north.

Tree dwellers need good eyesight in order to judge distances (The forest floor animals use hearing rather than sight as the tree

Left: A large number of plant species are present in just a small region of jungle. The dense undergrowth provides both food and shelter whilst sun-loving forms live high in the top canopy.

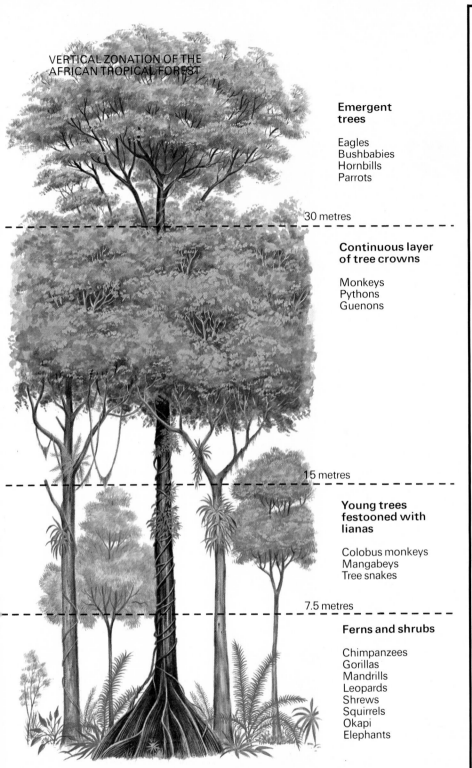

VERTICAL ZONATION OF THE
AFRICAN TROPICAL FOREST

**Emergent
trees**

Eagles
Bushbabies
Hornbills
Parrots

30 metres

**Continuous layer
of tree crowns**

Monkeys
Pythons
Guenons

15 metres

**Young trees
festooned with
lianas**

Colobus monkeys
Mangabeys
Tree snakes

7.5 metres

Ferns and shrubs

Chimpanzees
Gorillas
Mandrills
Leopards
Shrews
Squirrels
Okapi
Elephants

Climbers and fliers

Most jungle animals are climbers, like the numerous monkeys, rats, squirrels and snakes, or fliers like birds, bats and insects. Some animals move by gliding from tree to tree assisted by extensions of their body which act as parachutes. Thus there are gliding lizards, squirrels, phalangers, and even gliding snakes and frogs!

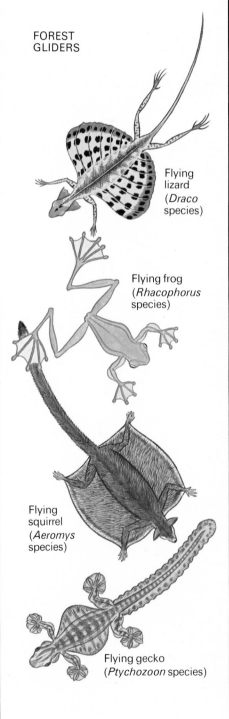

FOREST
GLIDERS

Flying
lizard
(*Draco*
species)

Flying frog
(*Rhacophorus*
species)

Flying
squirrel
(*Aeromys*
species)

Flying gecko
(*Ptychozoon* species)

trunks hinder good visibility.) Tree dwellers are also usually lighter in build than the heavier ground forms, and they have long, thin limbs and claws.

Rivers are the only open stretches to be found in the jungle, and they have a dense undergrowth along their banks. This provides food and cover for many animals, including some large forms which would find dense tree-covered areas difficult places in which to live. These animals include the rhinoceros of Java and Sumatra, the crocodile, the anaconda, the water buffalo and even the elephant.

19

Temperate forests

Above: *The mixed deciduous forests of the temperate zones provide a more open habitat than the rainforest, whilst still showing a large number of species and a range of sites in which to live.*

Between the tropics and the polar regions lie the temperate zones of the Earth. In the extreme north are the coniferous forests which border the tundra, and south of these are the temperate, deciduous forests, made up of numerous species of broad-leaved trees. The temperate forests of the southern hemisphere are found both in parts of Argentina and Chile in the Neotropical region, and scattered in narrow belts in New Zealand and parts of Australia.

The temperate forests are far more scattered than the more solid belts of coniferous or tropical forests. This is mainly due to man's agriculture and settlements. The soil of the deciduous forests is far richer than that of the coniferous forests, and the temperate climate is not as oppressive as the tropical one. Hence man has felled an enormous amount of the natural temperate forest. In North America only 1 per cent of the original forest remains, and that is concentrated mainly in park-land and the conservation areas.

One of the features of temperate forests is the autumn leaf fall with the canopy changing to shades of red and gold as the leaves die. Leaves accumulate and slowly start to decay during winter, the following seasons carrying the process of decomposition further.

The rich leaf litter teems with life – bacteria, fungi, worms and arthropods, all living on the rotting humus. These organisms in turn attract carnivores such as spiders, shrews, mice and centipedes. Tough leaves and twigs are eaten by millipedes and woodlice. Thus the litter layer is rich in life and produces minerals to encourage growth of the main layers of plants – the herbs, the bushes and finally the various species of tree.

The bushes and spreading crowns of the trees provide a rich source of food and shelter for a variety of birds, insects and small mammals. The common woodland

Above: *Damp areas of forest encourage the growth of mosses and liverworts where animals that require water can live. The pill woodlouse* (Armadillidium vulgare) *browses on old moss.*

insect-eaters such as hedgehogs, shrews, warblers, treecreepers, slow-worms and lizards, are all to be found. However, many insect-eaters of the deciduous woodland either hibernate, as do toads, hedgehogs and bats, or migrate as do many birds, to avoid the rigours of winter and the shortage of insect food.

Coniferous forests

The coniferous zone extends for 12,800 kilometres across the northern Holarctic. The commonest trees are pine, spruce and hemlock, with larch in the extreme north. These conifers produce dense shade under their canopies and the consequent shortage of light, plus the poor soil, restricts bushes and other shrubs growing on the forest floor. Only along the banks of rivers is there any tangled undergrowth, and other species of trees, notably birch, aspen and willow, form a thin belt flanking the banks and adding to the diet of the herbivores.

In some places the coniferous forests are very wet and marshy, like the muskeg of Canada and the taiga of Russia. These regions have numerous pools and waterways among clumps of conifers, providing a summer feeding ground for millions of wildfowl every year.

Below: *The pine forests of Northern Europe cover vast areas and help to retain and build up the soil on many rocky slopes.*

The conifers have survived in the cold, bleak north due to their leaves having a waterproof wax covering to prevent evaporation, plus a needle-like shape to lower wind resistance. The downward curving, flexible branches permit most of the snow to slide off and this prevents the snow building up to a great weight and perhaps tearing off branches.

Animals living in the northern forests have two problems – the cold of winter and the difficulty of getting food in winter. Any animals moving around in winter must have thick coats, and in addition many produce a coat colour which matches the countryside. The arctic hare, fox and stoat all have white fur in winter and this enables them to creep up on their prey unnoticed. Many rodents live underground with stores of seeds and roots, or even under the snow.

The conifers provide the food for most animals as there is so little undergrowth. Buds, bark, leaves and cones provide food for birds, squirrels and hares, which in turn are eaten by martens, owls, lynxes, wolves and bears.

Above: *Chipmunks are active members of the pinewoods throughout summer and autumn but hibernate in winter due to lack of food.*

The specialist eaters of cones and buds include red squirrels, chipmunks, crossbills, capercaillies and a whole host of insects such as pine-shoot-moths, sawflies and woodwasps.

Along the rivers running through coniferous forests live the larger herbivores such as moose and bears, as well as the active woodchucks and voles.

Grasslands

ideal feeding grounds for grazing animals and they are found in large numbers. The herds of antelope and bison benefit from the large numbers since they gain added protection from the vigilance of so many individuals. The lack of cover means that both herbivores and hunters must keep alert for, unlike the forest animals, they are exposed to view over long distances. This fact has led to the development of good eyesight in grassland animals. Many of them, such as the wild ass, the pronghorn antelope and the flightless emu, have also evolved fast locomotion over the flat plains.

Development of grassland depends on the climate and soil, and especially on the annual rainfall distribution. There are two main types of grassland – the temperate and the dryer savanna.

Above: *The magnificent sable antelopes* (Hippotragus niger) *live in groups, the darker males usually mating with 3 or 4 females. Their large ears are an important defence in detecting large predators.*

Temperate grasslands

These form a belt between the forests and the deserts, and often they are found in the interior regions of a continent where climatic conditions are suitable for their development. Temperate grasslands have a low rainfall and show regular seasonal temperature variations with hot dry summers and very cold winters.

Temperate grasslands are found in most of the continents and are represented by the rolling prairies of North America, the pampas of South America and the grassland ranges of Australia and New Zealand. However, by far the largest is the vast plain of the Eurasian steppes stretching 3200 kilometres from Hungary to China.

Unlike the various forests, temperate grassland always has a rich humus layer which forms, for instance, the famous 'black earth' of the Russian steppes. Although this soil is very fertile, the low annual rainfall prevents development of much more than grass and occasional scrub, except along the river banks.

The steppes and prairies are

Scrub grasslands

Where the rainfall is slightly heavier, for example on the slopes of mountains, or their protected hollows, scrub grassland is found. In these favoured spots it is possible for bushes to develop, providing extra food and shelter for different species of animals, such as small birds which eat grass seeds but, unless ground-nesting like the lark, would not otherwise have suitable nest sites. The species of scrub varies – eucalyptus in Australia and acacia thorn in Africa for example – but whatever the species, it is invariably tough and tolerant of having its buds either eaten or destroyed by frost.

Cultivated grassland

These may arise simply because native grasses have been replaced by a cereal crop, or they may develop where original forests have been removed to make way for livestock grazing, for fields or for plantations.

Whatever the history, the change represents the elimination of the natural animal life, although a new fauna may well move in to colonize man's agricultural land. Where the new plant life is restricted to only a few species, for example wheat on the prairies and sunflowers on the steppes, the new animal life will also be of a limited number of species. Hence the natural balance may well be unstable, for example there may be too many small herbivores. However, in most cases man's activities are maintaining this unnatural state.

Below: *This view of an acacia scrub grassland shows how tough the vegetation is for the wandering groups of herbivores. In the foreground young acacias can be seen growing through the dried grass tufts. This young growth is a favourite food source and much never survives to form trees.*

Mountains

The major mountain ranges of the world are the Rockies of North America, the Andes of South America, the Alps of Europe and, with their massive peaks, the Hindu Kush and Himalayas along the northern borders of India.

Although mountains cover only a small proportion of the Earth's surface, they are extremely important, biologically, for a number of reasons. Firstly, most habitats are controlled by the climate of the region, and this applies equally to mountain ranges. However, no other type of habitat has as dramatic an effect on the climate as does a mountain range, which greatly influences the weather in the surrounding areas. This is clearly shown by the Hima-

layas, which cause the southerly monsoon winds to rise when they reach the slopes and thus cause the rain to fall predominantly on the southern side of the range. The northern regions beyond the Himalayas are dry and desert-like. A second interesting feature is the difference in the climate, and hence the wildlife, on the two opposite sides of a mountain range. For example, the southern slopes of the Himalayas have a warmer, wetter climate than the northern slopes and hence support a richer plant and animal community.

The exact height at which any one type of vegetation or animal occurs will depend on the country and aspect. Thus the conifers of

Above: *This cold lake high up in the Andes has steep rocky sides and a stony bottom. The surrounding vegetation consists of short tough trees and shrubs capable of withstanding the wind and cold. In the distance are bleak snowclad slopes.*

Scandinavia and the northern Rockies are found from 300 to 1,500 metres, whereas mountains nearer the equator, such as the northern Andes, have similar forests extending several hundred metres higher. Similarly the two sides of the Caucasus show a difference—southern slopes having mixed forest above the level which, on the north-facing slope, may be almost bare of trees.

A further important point is the fact that mountain ranges may separate initially related groups of an animal species so effectively that after a long period of time, when mutations have occurred, the separated groups can no longer interbreed when brought together. Thus, they have become separate species as a result of their isolation from each other.

A journey from the gentle slopes at the base of a mountain range to the top provides an opportunity to see the different zones of vegetation, and hence the varying fauna. The plants at the base would be typical of the latitude, for example tropical forest on the equator and deciduous forest in the temperate zones, but in both cases these give way to coniferous forest as the altitude increases. The conifers eventually thin out and the timberline marks the upper limit for tree growth. Above this is an area of tough, wiry grass, with small bushes in protected hollows, and also a variety of colourful alpine plants such as gentians and alpine poppies.

Higher still the grasses are found only in crevices, and the exposed rocky slopes bear only mosses and lichens. These primitive plants and the tiny animals that live on them disappear as the permanent snow belt is reached.

Above this only the bleak, wind-swept, rocky outcrops remain free of snow and ice all year round.

The animals of the deciduous and coniferous forests have already been mentioned, but on mountain slopes these forests provide cover during winter gales for additional 'alpine' species, such as the chamois of Eurasia and the various bovids of the Himalayas, such as the goral and serow.

The alpine meadows or plateaux contain their own grazers. Tibetan gazelles and the Tibetan ass, yak, and chiru are all found in the Himalayas. In the Andes are found the chinchilla, guanaco and vicuna, while in the Rockies the bighorn and dall sheep live,

and in Europe the chamois and Alpine ibex graze on the tough vegetation. The carnivores are equally tough. The puma, snow leopard and wolf are the largest. The other hunters are the mountain birds of prey.

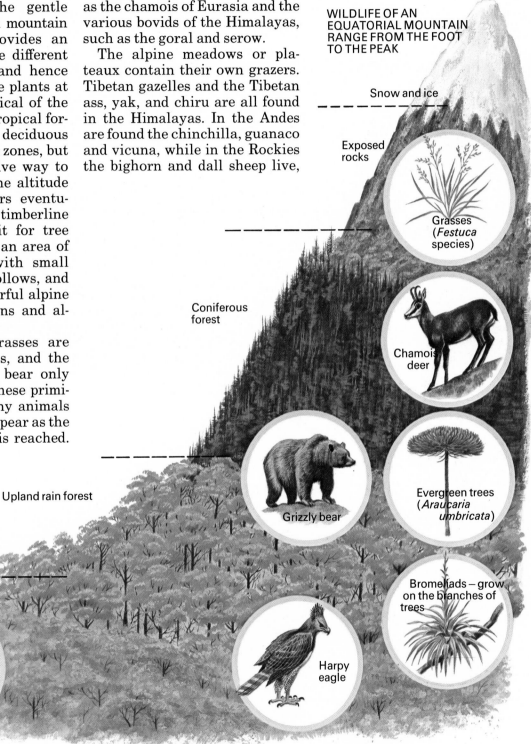

WILDLIFE OF AN EQUATORIAL MOUNTAIN RANGE FROM THE FOOT TO THE PEAK

Snow and ice

Exposed rocks

Grasses (*Festuca* species)

Coniferous forest

Chamois deer

Grizzly bear

Evergreen trees (*Araucaria umbricata*)

Upland rain forest

Bromeliads – grow on the branches of trees

Lowland rain forest

Harpy eagle

Bronzeback snake

Seas

The oceans cover just over 70 per cent of the earth's surface and provide a vast expanse in which life can exist. However, much of the ocean is unsuitable for any but the most specialized animals and plants because of the enormous depths which no light can ever penetrate. The average ocean depth is about 4,000 metres and the deepest trenches are approximately 11 kilometres deep.

The bottom of the ocean is as varied in its surface as is the land, with huge submerged mountain ranges that have peaks taller than Everest and valleys that descend for many kilometres. These great irregularities of the seabed are thought to have resulted from the movements of land masses during the formation and positioning of the continents.

All animals depend upon plants for their food, either directly as do the herbivores, or indirectly as do the carnivores. The green plants of the ocean are limited to the top hundred or so metres, where light can penetrate and enable the plants to make their food.

The most important plants of the sea are the minute drifting seaweeds – the phytoplankton. This phytoplankton provides a rich food for the mass of animal plankton (or zooplankton). The zooplankton consists of small shrimps, prawns, arrowworms and very small jellyfish. In addition, there are many animals that spend the young or larval stages of their life history drifting in the zooplankton. Such animals in-

Above: *The cool depths of the world's oceans teem with life. Shoals of fish, such as these herring, are common.*

clude worms, starfishes, sea-urchins, squid, octopuses and many other molluscs. Many small fishes also swim with the plankton, feeding on both types.

The great ocean currents, such as the Humbolt off South America and the Benguela off South Africa, bring a huge volume of water with fresh supplies of minerals and living organisms. These huge moving masses of water not only bring fresh supplies, but also help to mix up the waters of the oceans through which they pass.

This mixing of the waters and increase in the food supply causes a great burst of plant growth,

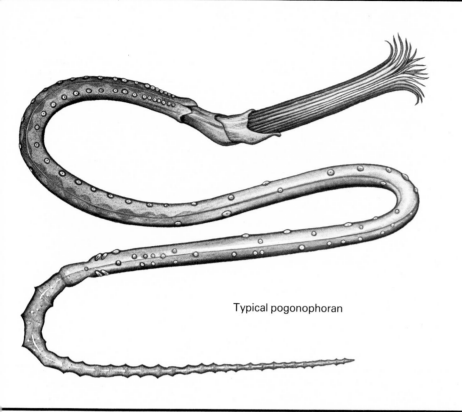

Typical pogonophoran

The ocean depths

The great dark depths of the ocean are known as the abyssal depths, and all the mineral supplies for this region come in the form of fragments of the plants and animals that lived somewhere above. This rain of debris provides a source of food for many strange animals. Curious stationary sea-lilies and starfishes live on top of this mud. Odd animals called pogonophora live buried in the silt with a fine tube-like crown of tentacles held above them. They have no gut but absorb the fine particles of food directly through the walls of the tentacles.

Many of the crustaceans have glowing patches on their bodies which possibly provide a means of keeping a group together when it moves through the black waters. Many of the fishes have on their bodies luminous regions which act as lures to other animals, and so provide the fish with a meal. As animal life is sparse in the abyss, many of the fishes there have huge jaws and stomachs, which enable them to eat a large meal when they do capture prey and so stay alive until the next successful catch.

followed by a similar increase in animal numbers. The zooplankton provides the food for slightly larger carnivores, notably free swimming fishes. These shoals of fishes are capable of deliberately moving from one area to another, unlike the drifting plankton. It is the movement of these shoals that is so important to man's fishing industry and hence we have learnt a great deal about fish movement and migration as a result of deep-sea fishing.

It is not only fishes that follow the plankton but also the great whales which feed on the drifting organisms. These whales are the whalebone whales which have specialized jaws to sieve off the zooplankton from the water. The blue whale (*Balaenoptera musculus*) can eat up to 2 tonnes of zooplankton in a day!

Right: *The source of food for all oceanic animals is the mass of fine plants — the phytoplankton. This is eaten by small animals such as crustaceans and their larvae. These animals make up the zooplankton, shown here.*

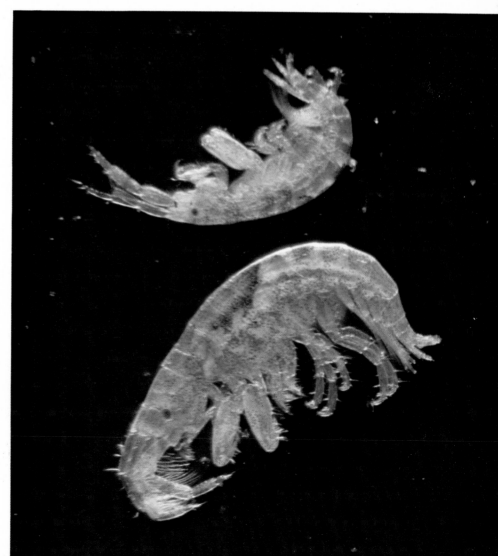

Coral colonies

Coral reefs and islands are formed from the chalky skeletons of millions of coral polyps, which are animals related to sea-anemones and living in huge colonies. Coral can only grow well in warm waters where there is plenty of light, which is needed by the green algae that live with the coral. These two requirements mean that coral structures are found mainly in the warm tropical seas. Fringing coral reefs exist near the shores. They grow on rocks a short distance off the coast. The barrier reefs are found some distance from the coast but are still bound to it. A coral atoll is formed around the edges of an erupted volcano and is built up from the seabed. As the volcano remains sink, so the coral grows, to stay in the warm, illuminated, surface layer. Eventually a ring of coral is left above the water, forming the atoll.

Coral reefs provide a honeycomb in which numerous brightly coloured fishes and shrimps live. The crown-of-thorns starfish (*Acanthaster planci*) has caused great damage in its attack on the Great Barrier Reef of Australia.

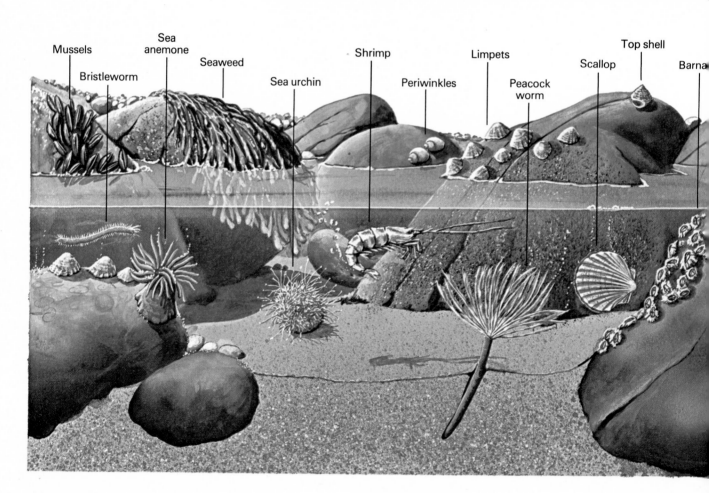

Mussels

Bristleworm

Sea anemone

Seaweed

Sea urchin

Shrimp

Periwinkles

Limpets

Peacock worm

Scallop

Top shell

Barna

Other mammals of the ocean are the seals, dolphins and killer-whales, all of which eat fishes, although the killer whales will also eat other marine mammals.

All the animals mentioned so far may be found in the illuminated layer of the sea, or the region immediately below. Other zones are recognized, namely the region where the slope of the continent continues out to sea, the continental shelf, and then the part where the shelf plunges towards the depths, the continental slope. Finally there are the great depths of the ocean which exist mainly below 1,850 metres.

The illuminated region covers part of the continental shelf and this region is very rich in animal life. Areas around Iceland and Greenland, and off the west coasts of Europe and South America, are some of the richest fishing grounds in the world.

The continental slope is often very irregular and has crevices and gorges through which mud flows move. Where the bottom is still, there is often a fine sediment of minerals that have washed down from the land and over the continental shelf. There is also a fine ooze of remains from the plankton. It is this 'rain' of dead remains that provides the food for the bottom-dwelling worms, crustaceans, starfishes and sea-lilies. These small invertebrates are in turn the food of squid, turtles and numerous fishes. It is these fishes that lure down the toothed whales that eat large prey, unlike the sieving whale-bone types.

The shores of the world provide a large range of habitats, each with its own fauna. All types, however, must have some way of overcoming the changes in humidity and temperature when the

Acorn barnacles
(Balanus balanoides)
feeding by means of feathery appendages which trap food

tide ebbs. Rocky shores have animals that are modified to withstand the pressure of waves, for example, limpets and barnacles. The sandy and muddy beaches have many burrowing worms, crabs and clams.

FAUNA OF THE SEASHORE

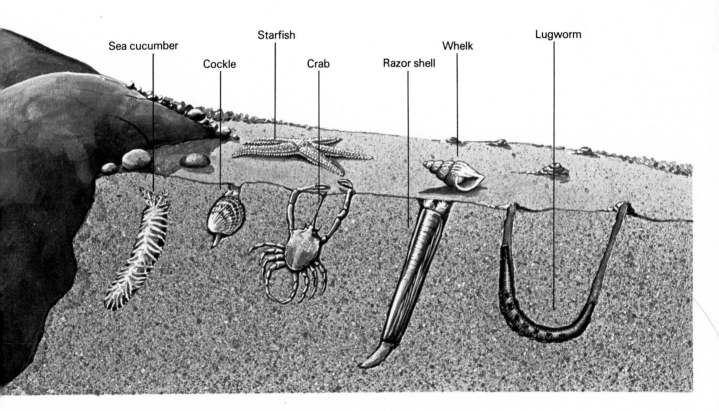

Sea cucumber Cockle Starfish Crab Razor shell Whelk Lugworm

Polar regions

These are the areas enclosed by the Arctic Circle in the north, and the Antarctic Circle in the south. The arctic polar region is a mass of ice floating in an ocean that is almost completely surrounded by land. The northernmost regions of the land masses of North America, Europe and Asia lie within the Arctic Circle and provide the areas in which polar animal life is most abundant. The arctic seas are frozen for all of the year but the northern tips of land that lie within the Arctic Circle do at least have a few weeks when the temperature is above freezing point.

The antarctic polar region is a continental land mass, greater in area than Europe or Australia, and it is surrounded by the cold southern oceans. The south polar region is much colder than its northern counterpart because of the vast expanse of snow and ice, some of which is over 1600 metres thick, and the freezing winds that blow outwards from the centre of this frozen continent.

The Arctic

Very little animal life can survive in the bleak open ice and snow of the central Arctic, but on the frozen land masses some animals can be found throughout the year, even during the dark cold winter months.

The lands that lie closest to the North Pole are called the *tundra* and here the wind is so cold and strong that no bushes or trees can survive, and so the vegetation consists of a mixture of lichens, moss, grass, sedge and tough

Above: *The tundra is bleak and open, the rocks being worn and splintered by frost action. Vegetation grows mainly in hollows between the sheltering rocks and helps to bind the loose rock chips together. Flooded regions provide homes for crustaceans and migrating birds.*

flowering plants, such as the arctic heathers. During the winter months the musk ox, the caribou or reindeer, and the lemming dig away the snow to uncover the lichens and mosses which they need to survive.

These tough grazing animals provide food for such predators as the arctic fox and snowy owl, both of which eat the lemming, and the polar bear, which is capable of killing a solitary musk ox or caribou. The polar bear is the largest carnivore in the Arctic. It is a great traveller, being expert at swimming and climbing over

Right: *Adelie penguins* (Pygoscelis
deliae) *are the most widely distributed
of the antarctic penguins and build nests
of stones. To reach stones they often
cross 80 kilometres of ice.*

the ice and snow. Fishes and seals
that flourish in the Arctic Ocean
provide most of the bear's food.

With the coming of spring and
summer the snow and top few
centimetres of soil thaw and
cause shallow ponds and streams
to form over the entire tundra
region. In this surface water crus-
aceans, frogs and insects thrive
and reproduce, whilst the tough
arctic plants quickly produce
leaves and flowers before the cold
returns. This sudden burst of life
is very attractive to birds, and the
polar regions provide a major
summer retreat for thousands of
birds such as mallard, teal, eider
duck, snow goose, swan and hundreds of other species. Here the
birds can feed on the abundant
pond life and breed in compara-
tive safety, away from man and
other predators.

Below: *The snowy owls* (Nyctea
candiaca) *live in the Arctic feeding on
rodents. This immature specimen shows
the talons that will soon be hidden by the
feathery leg covers.*

The Antarctic

In contrast to the arctic tundra,
there are only 2 species of flowering plant found in the Antarctic
and the rest of the vegetation
consists of lichens, algae and
mosses. Very few invertebrate
animals are found; just a few
kinds of crustaceans and insects.

There are only 8 species of
animal that breed around the
edges of the Antarctic — 4 species
of seal, 2 species of penguin, the
southern polar skua and the snow
petrel. All of these rely for food on
the abundant animal life in the
ocean, although the skua will
also take young adelie penguin
chicks if they are left unguarded.

Arid regions

Savanna

In the earlier section on the grasslands, the prairies, steppes and pampas, a brief mention was made of the savanna. Savanna grassland exists in the tropics where the annual rainfall is restricted to a few summer months, although the amount of rain falling may well be greater than that falling on some temperate grasslands. At the present time these conditions are found in 4 areas – Africa, northern Australia, northern South America and parts of India. It should be emphasized that the savanna of India is not of natural origin but is the result of man clearing the forests that once existed there.

The savanna of South America is found on either side of the Amazonian forests, but is mainly concentrated in that area of Brazil to the north of the vast pampas.

Most studies of savanna have been carried out in Africa where over 35 per cent of the continent is covered in the tall grasses that are so typical of this form of grassland. Like the savanna of South America, that of Africa is divided into two by a belt of tropical forest.

The African savanna has some of the finest scenes of large grazing mammals to be found anywhere in the world. A visitor is immediately struck by the size of the herds and the vast numbers of different animals present. The gazelle, zebra and hartebeest move slowly along, eating steadily, whilst always remaining alert for any suspicious movements or scents of their enemies – the lion, leopard and cheetah.

Despite the numbers of differ-ent species involved there is little direct competition between the herbivores, as each type tends to specialize in a particular form of vegetation. Thus, for example, the giraffe feeds on thorn trees up to 5.5 metres high, while the eland feeds on the buds and foliage of the bushes and the gazelle eats grasses. The numerous rodents

A giraffe feeding on a thorn tree

feed on roots, seeds and leaves at many different levels. Birds also vary in their diet, some eating seeds, others eating insects and small grubs, and the oxpecker devouring parasites removed from rhinoceroses and buffaloes.

The vast number of mammals and birds are accompanied by even more insects which live on the vegetation, other animals, and the droppings and dead bodies of the larger beasts. This wealth of insects is attended by small carnivorous animals like scorpions, spiders and centipedes and these in turn fall prey to the numerous lizards, snakes and rodents. Thus at many levels animal life relies, either directly or indirectly, on the tall grasses, the thorn bushes and the flat-topped acacia trees.

The South American savanna resembles very clearly that of Africa, although the grass here is often shorter. However, in some northern parts, for example the Llanos of the Orinoco plains, the countryside is far more open and desolate with only scrub for shelter, the trees being restricted to the river banks and the swamp lands closer to the equatorial forest. The savanna south of the Amazon is very similar.

These conditions have proved more suitable for the evolution of smaller herbivores than those prevailing in Africa. Thus there are not as many deer in the South American savanna as in the African, the most important being the pampas deer, but there is a greater number of smaller mammals and insects. Rodents are found throughout the savanna and their range extends into the pampas. The capybara, Brazilian cavy and viscacha have been mentioned in connection with the pampas, but in the savanna they are joined by many others. South American rodents are kept in check by such carnivores as the pampas cat, the skunks, and numerous species of fox and wolf and by owls, falcons and other birds of prey.

Right: *These zebra have probably covered long distances to reach this water hole, the sparse vegetation showing the effects of a long drought. The acacia scrub and trees are the only large plants growing in the vast area of dead and dried grass.*

Desert regions

About one fifth of the surface of the Earth is covered by some type of desert, and most of these occur in the Holarctic region. The largest desert is the Sahara in Africa, consisting of 9 million square kilometres of dry rock, sand and mountain. Other deserts are found in North America, central Asia, south-west Africa, Australia and south-west America.

The deserts in the centre of a continent, for example the Gobi Desert in Asia, form because of their distance from the moist coastal winds. Others, such as the North American deserts, are formed in the lee of mountain ranges, all the rain falling on the windward side. As a result of their altitude the Tibetan and Gobi Deserts are cold regions, whilst others, for example the Sahara and Arabian Deserts, are very hot.

Desert rainfall is very low, often less than 25 centimetres per year, and this means that very few plants, other than tough grasses and a few specialized plants such as cacti, can grow. Such limited vegetation means that animal life is restricted by a shortage of food. The shifting top soil, resulting

Above: *The Namib desert shows some of the problems of desert animals. There is little vegetation and the ground is either splintered bare rock or sandy gravel. The lack of shade makes life almost intolerable.*

from a lack of plant cover, and the problems of shortage of water and excessive heat (or cold) also limit the animal life.

Burrowing places an animal in the shade, and the difference in temperature between scorching sand on the surface, and that only a short way underground is quite remarkable. Measurements in rodent burrows in the Mojave Desert gave a difference of over 35°C between the opening and the

Right: *The kangaroo rats are some of the most successful colonizers of North American deserts. They even live in Death Valley, California.*

end of the tunnel 50 centimetres underground.

Many animals aestivate, that is they become dormant during the hottest time of the year. One animal that does this is the Mojave ground squirrel.

The insects, scorpions and spiders have adapted very well to life in the desert, as have certain lizards, in particular skinks and geckos, and some snakes. However, even these forms usually shelter from the heat by burrowing or hiding under stones or in crevices in the rocks. The commonest mammals are the rodents, such as the kangaroo rat of North America, the jerboa of North Africa and the many ground squirrels of North and South America. Other desert mammals include the golden mole of south-west Africa, the Sahara fox and the famous Arabian and bactrian camels. There are few birds in the central desert, although many live in the outer fringes, or near oases, and because they can fly, they can cover very large areas, which may include the dry central regions. Examples of desert birds are stone curlews, coursers, elf owls and desert quails.

Living without water

The problem of water shortage is overcome in a number of ways by different animals. Some, such as the North American kangaroo rat (*Dipodomys deserti*), can exist without taking in water. This particular small rodent gains its water from the fats in its food, which consists of dry vegetation and seeds.

Many rodents and insects can produce very concentrated urine and most of the desert mammals are incapable of sweating. The sand grouse (*Syrrhaptes paradoxus*) soaks up water in its breast plumage, from the occasional water hole, and then carries it back to its young which are thus protected from the scorching sun by their parents' damp bodies. The thorny devil (*Moloch horridus*), a spiny Australian lizard, absorbs water from damp sand via its specialized scales.

Right: *The sandgrouse shown is one of 16 species which live in semidesert or arid regions. They never perch in trees, many resting from the heat of the day amidst bushes. Adults will fly hundreds of kilometres each day to drink at communal watering places.*

Freshwater environment

Freshwater covers only a small proportion of the Earth and accounts for only one third of 1 per cent of the water on the planet. Nonetheless it provides a great variety of habitats that almost equal in number and range those provided by the land regions.

Many great rivers start their life high in the mountains as a result of the outflow from a mountain spring and the rainfall. Mountain streams are very fast flowing and usually shallow with rocky bottoms. The amount of water flowing varies with the rainfall, but, when flooding down its narrow channel, it is capable of wearing away the crumbling, unstable banks and carrying rocks and boulders in its current. Few plants can live in the shifting, stony bed of the stream, and the only aquatic plants are mosses and liverworts with the occasional hardy fern overhanging the surface.

As the mountain stream becomes a little wider a few animals appear, all of them being modified for life in this cold, bubbling water. These include animals which live in the crevices of rocks, such as freshwater shrimps (*Gammarus*). Others are flattened and have claws or suckers with which to cling to rocks or stones. For example, stonefly and mayfly nymphs have claws, whilst limpets and leeches use their suckers. The only fishes are strong swimmers, such as the brown trout and stone loach, that can head into the current.

As the stream makes its way out of the foothills of the mountains it becomes wider and slower moving and also richer in minerals, due to draining from the land on either side. Plant life increases and the bottom is now covered in silt, sand and mud, so that the animal life includes burrowing forms such as various worms and molluscs. The growing number of plants provides homes for numerous insects and their larvae, such as alderflies, mayflies and many different kinds of beetles. Fish life is richer, with minnows, eels and young salmon

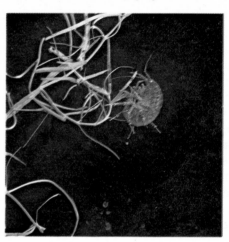

Below: *There are many different species of freshwater shrimps and* Gammarus *is common in European waters. Like many species it swims on its side between the weeds and stones scavenging for food.*

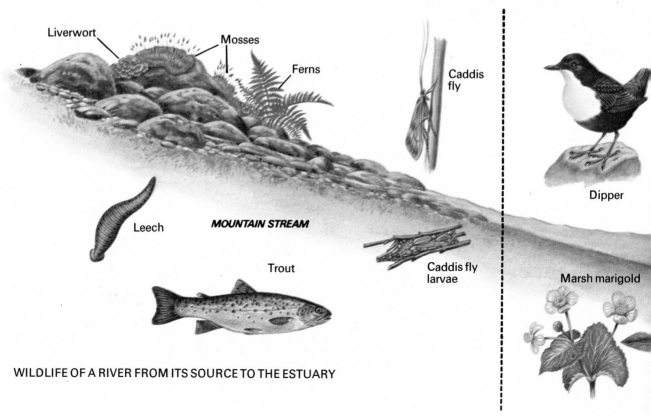

Liverwort

Mosses

Ferns

Caddis fly

Dipper

Leech

MOUNTAIN STREAM

Trout

Caddis fly larvae

Marsh marigold

WILDLIFE OF A RIVER FROM ITS SOURCE TO THE ESTUARY

fry. Birds are more numerous. They feed on the insects alongside the river and even in its water. The dipper actually walks along the bottom.

When a river reaches the flatlands it winds more frequently as it drains into lakes or the sea. This is the region with the densest plant and animal life. Reeds and rushes fringe the edges of rivers and lakes, providing homes for insects, birds, and fishes such as pike and perch. Finally mammals, including voles, otters and shrews, live on the river banks. In the tropics, dense jungle grows to the water's edge and the river provides an easier means of movement for animals than the dense undergrowth. Hence larger animals such as the tapir, caiman, elephant and deer may be found. In Eurasia the mammals and reptiles are smaller. The birds form a dramatic part of the wild life with heron, duck, grebe and kingfisher moving through the waters in search of food.

Lakes vary in animal life, depending on the age of the lake and

on the country in which it is situated. Young lakes still have stone bottoms, few minerals and little plant and animal life. The older ones have partially silted up and teem with life suited to waters of that region. Thus African lakes draw hippo, deer and flamingoes in great numbers, and also the hunters that accompany such herbivores.

Above: *This Kenyan lake is rich in different plant species, floating or submerged, providing food and cover for many animals.*

Where the water table is high and drainage poor, marshes and swamps are found with their dense sedge and reed beds. Here animals such as warblers, frogs and terrapins all thrive on the abundant insect life.

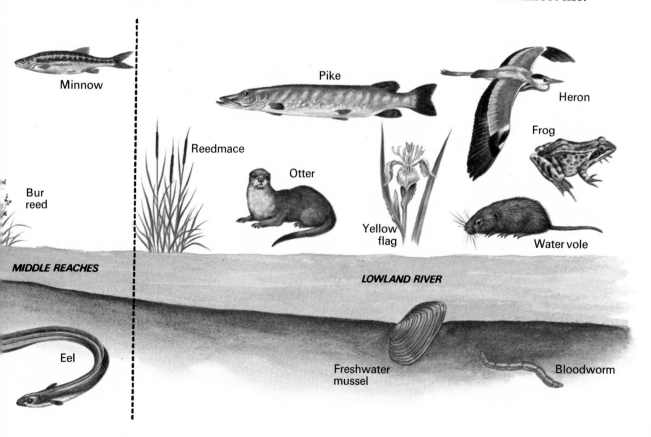

Minnow

Pike

Heron

Reedmace

Frog

Bur reed

Otter

Yellow flag

Water vole

MIDDLE REACHES

LOWLAND RIVER

Eel

Freshwater mussel

Bloodworm

Special environments

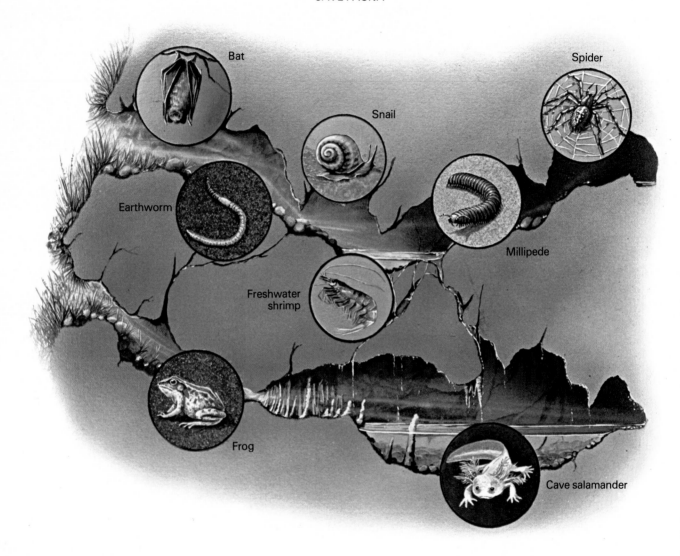

Bat

Snail

Spider

Earthworm

Millipede

Freshwater shrimp

Frog

Cave salamander

In the preceding pages a variety of habitats have been described with examples of the different inhabitants of these regions. All the various habitats described so far are relatively common; but in this section we shall look at some of the less usual habitats which support a rather more specialized fauna. Some of these are natural, for example caves and boiling mud springs, whereas others, like aquaria and nature reserves, are creations of man.

Caves

Caves are found in two places – around the shores of countries where the sea has destroyed part of the cliffs, and inland in chalky or limestone country, for example Cheddar in Britain, parts of Arkansas in the United States and inland regions of Yugoslavia. Marine caves are merely extensions of the upper, or splash zone of the beach and so we will not be concerned with them here.

Limestone caves are formed by acidic water draining through the rock and creating huge caverns. Many of these underground vaults have lakes and streams flowing through them.

and these unlikely bodies of water form the home of some strange animals.

As there is no light deep underground there cannot be any green plant life to support the animals. Therefore all food must originate outside; it may take the form of vegetation that has blown in, or minerals and rotting vegetation that have been washed in by flood waters or possibly materials that have been carried in by animals such as cave bats or birds. The droppings of the animals assist the growth of bacteria, fungi and small unicellular animals and it is this simple life that is the start of all food chains.

Small invertebrate animals such as worms, mites and a few species of insect live on the bottom and around the edges of the waters. Swimming freely in the water are larger crustaceans such as the blind shrimps of the North American caves.

The carnivores are represented by a number of small fishes and by strange salamanders with spindly legs and flattened bill-like mouths. These salamanders are found in North American caves as well as in some European caverns, such as the huge Yugoslavian caves.

In deep caves that have connections with the outside, live the more advanced animals such as the bats which roost in the day

and emerge at night, and some of the less common birds such as the oilbird of South America. The oilbird (*Steatornis caripensis*) produces a series of squeaks and clicks to aid its navigation in the dark – a very similar process to the high pitched squeaks of the bat 'radar' system. Caves that are inhabited by bats have a large

Above: *Oilbirds spend the day in caves roosting on ledges and emerge at night to feed. Their nests are built from regurgitated fruit juices and droppings. Natives collect the nestlings for their blubber.*

insect population which lives on the droppings and on the dead bats. These in turn attract other predators such as spiders and rodents.

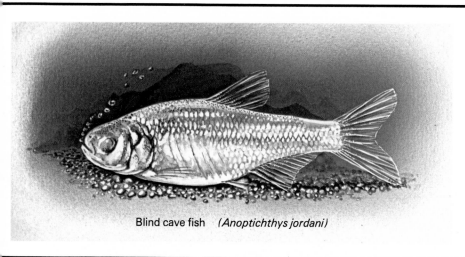

Blind cave fish   (*Anoptichthys jordani*)

Cave specialists

All the cave animals so far described have been isolated from the rest of the world in the dark caverns and have evolved along separate pathways from their relatives in the world above. Most are blind, since light is absent, and to make up for this many have developed specialized organs sensitive to touch. For example, there are the tentacle-like barbels of the African cave catfish, the delicate antennae of the shrimps and the pressure sensitive devices of the tiny fishes.

Man's lodgers

Human dwellings contain many animals in addition to their rightful owners. Obviously there are many that can be considered as 'tourists'; they are often found but are merely passing through. Butterflies, wasps, and mosquitoes come into this category. Other animals move into man's homes and establish themselves, and these can often be troublesome. Kitchens and rooms containing food are an obvious attraction for scavengers, and hence cockroaches and flies, for example, will certainly breed and raise offspring generation after generation unless kept in check. Other unpleasant scavengers are mice and rats which not only eat food and carry disease but cause great damage to structural features such as furniture, woodwork and even, in the case of rats, to electrical wiring and plumbing.

Beetles make use of clothing, furniture, carpets and timber for their larvae and they can cause the fabric of a building to become unsafe. In tropical countries the termite is even more devastating in its attack on house timbers.

Papers, books and wallpaper provide homes for the tiny silverfish, a primitive wingless insect, and the booklouse, which seem to prefer the glue in the bindings of books. Both these animals need damp conditions and hence are found in cellars or, in the case of the silverfish, bathrooms.

The thatch on some houses can provide a home for numerous beetles, sparrows and rodents, all of which aid the fungi in the slow destruction of this covering material.

The numerous insects found in houses attract hunters, notably the spiders. The common European house spider (*Tegenaria*) is found in many countries of the world and is common throughout

Above: *Cockroaches are amongst the most common scavengers for food in shops, homes and warehouses.*

the Holarctic region. Its web is an untidy mass of threads spun across corners of window frames, doors and cupboards, and it may become dusty. Without the many species of spider living in our homes we would find our lives even more difficult as a result of invasion and damage caused by insects.

In tropical houses, insects are even more numerous, although their food sites are similar to those of the temperate zone. However, the range of carnivores is increased, and numerous snakes and lizards live in the eaves, thatch and rafters to descend whenever a moth, beetle or cockroach comes near. Animals such as the mongoose are encouraged as they not only keep down potentially dangerous snakes, but are also fierce predators of the mice and rats.

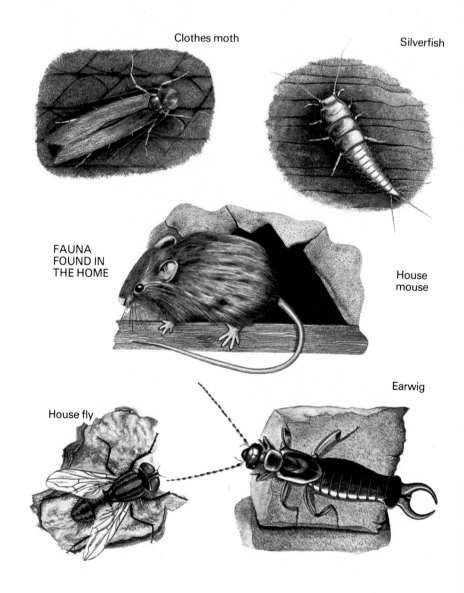

Clothes moth

Silverfish

FAUNA FOUND IN THE HOME

House mouse

Earwig

House fly

Animals enclosed

Artificial enclosures for animals serve a number of purposes. In the past, aquaria and zoological gardens were simply places of entertainment for the public, places where they could marvel at the various animals. Today zoos and safari parks have an important function in providing protected and controlled areas where animals may breed successfully, and in so doing help to overcome the ravages that have occurred in the wild. Too often animals have been, and are still being, displaced by the growth of human populations or are killed for their hide, tusks or feathers.

A similar role is performed by the larger nature reserves which are areas of natural country preserved for the native animals that are protected by law from attack by man. The huge reserves of Africa and North America do a splendid job and also permit large numbers of people to study animals behaving naturally. It is the retention of the natural vegetation that makes these areas so valuable, since the food cycles can proceed in a normal fashion and thus the unnatural aspects of life in the parks and zoos are overcome. Animals that have been saved from extinction by breeding in captivity include such species as Père David's deer (*Elaphurus davidianus*), which has now been reintroduced to China after being exterminated there, and the Hawaiian goose or néné. The Hawaiian goose population was reduced to only 42 individuals but the number was built up at the Wildfowl Trust at Slimbridge in England from a

Above: *Père David's deer was saved from extinction at Woburn in Britain. This beautiful animal was wiped out in its native China by 1900. For once man has saved an animal species.*

small group of 2 females and 1 male. The breeding programme was so successful that these birds have been reintroduced to their native islands.

Aquaria and vivaria provide means of studying fishes, amphibians and reptiles in regions of the world where the natural climate may be quite unsuitable. It is relatively easy to control the temperature and humidity in glass tanks and hence many of these vertebrates can be studied in this fashion. However, it must be remembered that, although they may live and breed, their behaviour patterns may not always be the same as occur in the wild.

Parasites

These are organisms that live on or in another animal or plant and cause it harm. Fleas and ticks are common parasites on the skin of our domestic animals, such as cats and dogs, whilst many of man's most important agricultural animals are parasitized by internal parasites such as tapeworms, hookworms and the various blood parasites.

Many animals live in close association with each other and cause each other no harm at all. For example, many worms that live in mud and silt at the edge of the sea share their burrows with several other animals. The ragworm (*Nereis fucata*) is often found living within the shell of the hermit crab (*Pagurus bernhardus*), without much mutual influence, and this is known as a *commensal* association.

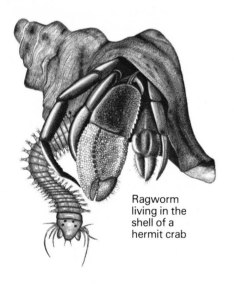

Ragworm living in the shell of a hermit crab

Some parasites only briefly visit their hosts. Mosquitoes and leeches, for example, take a meal of blood and then leave the host. These parasites are still capable of moving in a fashion that is normal for their group, and they differ from their free-living, non-parasitic relatives only in their diet.

Left: *Fleas are very successful parasites as they have colonized every species of mammal in the world. They possess powerful legs for leaping, hooked claws for gripping and narrow bodies to slide between hairs.*

The more specialized parasites have sacrificed their powers of locomotion and tend to stay with the source of food, the host, more or less permanently. Ectoparasites, parasites that live on the outside of their host, need some means of staying on the host. Fleas and lice have developed hooked limbs and biting mouthparts to this end, while some parasitic flatworms have developed suckers. Fish flukes for example stick on the gills of fishes.

Internal parasites, or endoparasites, have additional problems. Not only must they stay attached to their host, but they have the problem of getting there in the first place, and then stopping the host from getting rid of them once they have arrived. Tapeworms enter their hosts in infected meat and, after passing through the host's stomach, attach themselves to the lining of the gut by means of rows of hooks and suckers.

Some parasites have more than one host in the course of their lives. Thus the liver fluke of sheep is found in a freshwater snail as well as in the sheep, and the human blood fluke, causing bilharziasis, also makes use of a snail.

Parasites of man, such as those causing bilharziasis and malaria, are poor parasites from the point of view of Nature since they kill their hosts relatively quickly. It is better for the parasite if the host can be kept alive to provide a permanent food supply and thus ensure the parasite's successful reproduction.

Left: *This freshwater leech has attached itself, by means of the sucker at its head end, to the tail of a fish.*

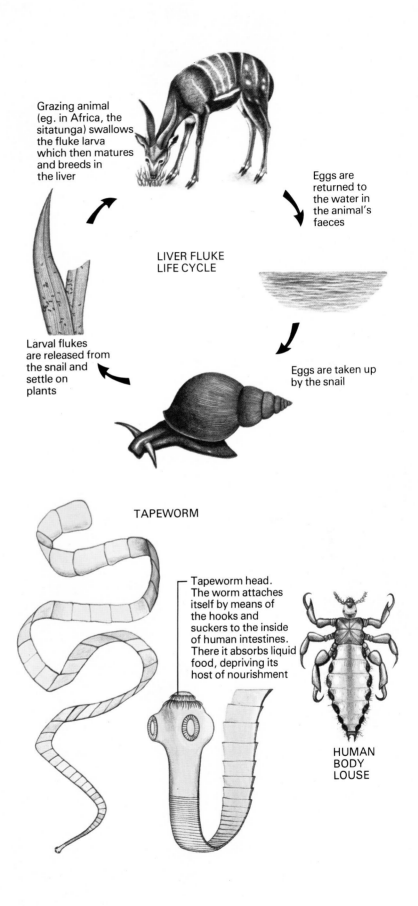

Grazing animal (eg. in Africa, the sitatunga) swallows the fluke larva which then matures and breeds in the liver

Eggs are returned to the water in the animal's faeces

LIVER FLUKE LIFE CYCLE

Eggs are taken up by the snail

Larval flukes are released from the snail and settle on plants

TAPEWORM

Tapeworm head. The worm attaches itself by means of the hooks and suckers to the inside of human intestines. There it absorbs liquid food, depriving its host of nourishment

HUMAN BODY LOUSE

Migration

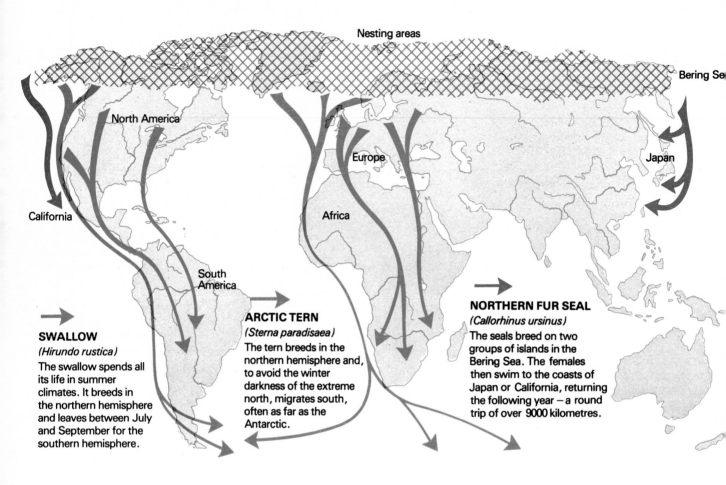

Nesting areas

Bering Se

North America

Europe

Japan

California

Africa

South America

SWALLOW
(Hirundo rustica)
The swallow spends all its life in summer climates. It breeds in the northern hemisphere and leaves between July and September for the southern hemisphere.

ARCTIC TERN
(Sterna paradisaea)
The tern breeds in the northern hemisphere and, to avoid the winter darkness of the extreme north, migrates south, often as far as the Antarctic.

NORTHERN FUR SEAL
(Callorhinus ursinus)
The seals breed on two groups of islands in the Bering Sea. The females then swim to the coasts of Japan or California, returning the following year – a round trip of over 9000 kilometres.

Migration is the term used to describe the regular movement of animals to new areas from which they subsequently return. These journeys take place at set times of the year and are therefore seasonal movements. Some reasons for these movements are changes in the weather, changes in the food supply and the pressure of numbers. For example, the Arctic is far more hospitable in June than it is in January, when the temperature is extremely low and the wind is bitterly cold. To these climatic problems must also be added the lack of food for both herbivores and carnivores. These problems are overcome by such animals as birds which fly south during the arctic winters to return only when the conditions have improved, that is when the temperature has risen, the snows have thawed and the plant and insect life is thriving once more. Under these conditions the newly returned migrants find the Arctic a good place in which to feed and also to breed. The fact that very few animals live permanently in the Arctic means that there is space in which to claim a breeding territory and this provides a further reason for the northward migration.

Birds

Birds are the greatest migrants in the animal kingdom, as their powers of flight enable them to cross rivers, mountains and even oceans during their journeys. Arctic terns (*Sterna paradisaea*) breed in the extreme north of Europe, Asia and America. In autumn they fly south all the way to the southernmost tips of Africa, America and Australia. Here they stay until the following February or March, when once again they return to their northern breeding ground. These incredible journeys cover over 35,500 kilometres in one year.

Fishes

After the birds, fishes are the next greatest migrants. Generally they

Right: *This flock of arctic terns* (Sterna paradisaea) *are passing along a shore where they will feed. Typically they eat small fish, molluscs and crustaceans but they will also eat insects. Narrow wings and tapered body aid fast flight.*

spend much of their time on the move, but this is connected with feeding and drifting in the water currents, and is not true migration.

The salmon is a well-known fish of Europe and North America and has great importance as food for man and other animals. The adult salmon live in the sea but return to the fresh water of the rivers to breed. During these journeys upriver they have to make tremendous efforts, clearing, for instance, small waterfalls with leaps of 3 to 4.5 metres. After breeding, the adult salmon drift downstream and usually die, exhausted by their journeys upstream against the current. The young salmon hatch and slowly make their way, with the help of the current, down to the sea. This journey to the sea may take up to 4 years. On reaching the ocean, the young salmon disperse in all directions and do not return until they are full grown. Eventually, however, the salmon will return to the mouth of the river in which they started life. They then must begin the tremendous struggle against the current to reach the breeding grounds in one of the river's tributaries.

Mammals

Most mammals are land-dwelling and are therefore restricted from moving great distances by rivers and mountains. In Africa it is still possible to see herds of wildebeest, zebra, elephant and Thomson's gazelle making their steady movements across the dusty plains in search of fresh green vegetation. It is the problem of finding fresh food, and water in

the dry season, that keeps the herds moving. When the rainy season starts, the parched dusty plains soak up the water and fresh grass appears. This supply of new vegetation comes just as the young are born and hence provides them with a good start in life. After a few weeks this food is

eaten and the herd must move on, but the young, by now, are strong enough to keep up with the adults.

Below: *This mature salmon shows the power and grace for which they are famed. The powerful tail muscles and broad fin drive its body clear of the torrent and rocks beneath.*

Animal behaviour

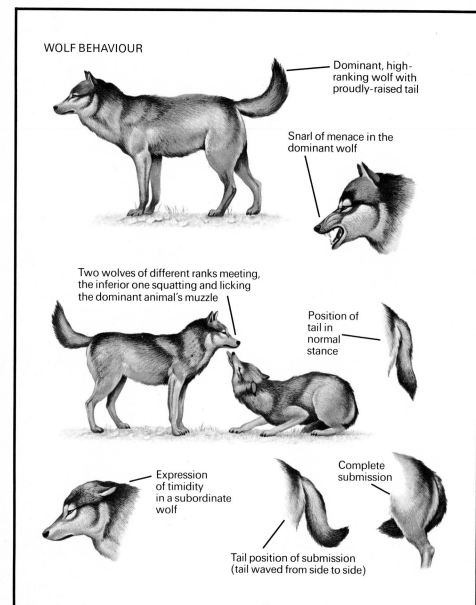

WOLF BEHAVIOUR

Dominant, high-ranking wolf with proudly-raised tail

Snarl of menace in the dominant wolf

Two wolves of different ranks meeting, the inferior one squatting and licking the dominant animal's muzzle

Position of tail in normal stance

Expression of timidity in a subordinate wolf

Complete submission

Tail position of submission (tail waved from side to side)

Body language

Spanish scientists studying wolves in a large European reserve found that the position of the ears, the angle of the tail and the amount of teeth displayed were all very important in the life of the pack wolf. The leader of the pack was the only wolf allowed to have his tail raised above the level of the back, and any other wolf that dared to wag its tail in an upright position was treated as a challenger for leadership.

Thus wolves, like chimps, use their body to tell their companions about their feelings, and also to advertize their position in the life of their society. Man still uses signs to tell others of his social position, but instead of body movements, he now relies mainly on his manner of speaking to impress others with his knowledge. He also uses his possessions as symbols of his status. Many people, for example, care for their front gardens, not simply because they like flowers or a neat front to their house, but because they wish to impress passers-by. Similarly the latest car or foreign holiday can be used to tell others about a person's status — a human equivalent to the raised tail of a wolf.

Communication

In an experiment undertaken in the United States, a scientist raised a 7.5 month old female chimpanzee alongside his own 10 month old son. The two youngsters were treated in the same way at all times, and in the early stages of the experiment it soon became clear that the chimp was responding better than the little boy. She learned to obey simple instructions and became far better at handling and moving objects. However, as time passed the young boy progressed from scribbling happily with his monkey friend, to writing letters and drawing shapes that were recognizable as household objects. Also his nonsense sounds and cries developed into words, which later he joined up to form phrases. These two changes, from scribbling and garbled sounds to writing and speech, were never achieved by the chimp.

Learning to write, read and speak allowed the boy to understand others and to make people aware of his needs: without these abilities his level of attainment would have been only a little above the chimp's.

Man is a very intelligent animal with an extremely complicated way of life. He has evolved a number of ways of telling others what he means: these include speech, writing or possibly gestures. Man's spoken and written history has enabled knowledge to be passed from one generation to another. This has obviously been of great benefit to him. Modern man also uses records, tapes and films in his communications with others.

The chimp (*Pan troglodytes*) is also a very intelligent animal and leads a complicated life when compared to many lower animals. However the chimp uses only its hands, feet, lips and teeth in moving objects. Furthermore it does not have a very large number of sounds with which to form a language. Chimps, like many mammals, can inform each other of their feelings by using facial expressions, and scientists have recorded on film the critical movements of their lips, cheeks and eyes.

FACIAL
EXPRESSIONS
IN THE
CHIMPANZEE

Pleasure
(smiling)

Thoughtful

Annoyance
(appears to
be laughing)

Territory

So far only mammal behaviour has been mentioned and much of it human. Therefore it would be worth considering animals generally and to think of the reasons why animals communicate with each other.

Firstly there is the necessity to claim a territory, since many animals remain within a given area, providing that it can pro-

duce enough food. Birds very often advertize their presence by perching on a post or tree and singing loudly. Usually this display is effected by the male bird and serves to warn other males that the particular area is claimed. The European robin (*Erithacus rubecula*) provides a very good example. This aggressive little bird warns off his rivals by singing and thereby saves himself the bother of actually fighting for his territory.

Many mammals mark out their territory. The hippopotamus, for example, deposits a mixture of dung and urine along the edge of its grazing area. Deer often use the pungent fluid from special glands, located near their eyes, to mark twigs, bushes and even the ground, in an area over which they wish to assert their authority. Thus these mammals both communicate by means of scent, while the birds do so by means of sound. The male robin is roused by the colour of the red breast feathers of a challenger and thus sight is also an important aspect of signalling.

Below: *This male robin is displaying the brightly coloured breast feathers and also sings to show his territorial claims.*

Mating behaviour

Territorial claims are made not only for the individual animal's benefit. Male animals may secure a site in order to attract females so that mating and raising of young can occur. Obviously it is better for the species if only the strongest males succeed in reproducing, and so there is great competition for the females. However, a fight does not usually continue until the death of one of the combatants since this is wasteful. It is better if the loser surrenders and then leaves for less populated regions, thus spreading the species. There comes a point in any fight, especially between males, where one of the fighters realizes that he is going to lose and hence does not want to fight on. What usually happens then is that this animal adopts a special stance that tells the stronger opponent that he has won and that the fight is over. In wolves this surrender posture involves the loser rolling on its side or back to expose the belly and throat – two regions that are very vulnerable and that normally are carefully protected.

Parents and young

Another reason for communication is to ask for something or to seek help. The wails of a hungry baby are very effective in gaining attention. Baby birds encourage their parents to feed them by producing visual signals. For example song thrush nestlings have coloured linings to their mouths

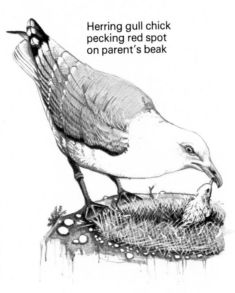

Herring gull chick pecking red spot on parent's beak

Above: *Two red deer stags* (Cervus elaphus) *are fighting with their interlocked antlers. In the background a female grazes.*

and the parent birds will always try to feed gaping beaks displaying such colour, whether genuine or man-made. Herring gull chicks demand that their parents feed them by pecking the end of the parent's bill, where there is a red spot. Thus the parent presents a coloured spot signal to the young, and the young show that they wish to be fed by tapping. Hence two different methods of communication are displayed in one species (*Larus argentatus*).

Finally of great importance to social animals is the signal that indicates an individual's position and thus helps the group to stay together for protection. This signal may be one of scent, as in the case of some fishes and moths, or one of sound as produced by many insects. Alternatively, the signal may be a flash of coloured fur, scales or feathers.

Right: *The dunnock chicks* (Prunella modularis) *show their coloured mouth linings. The parents respond instinctively by feeding them.*

48

Reproduction in the animal world

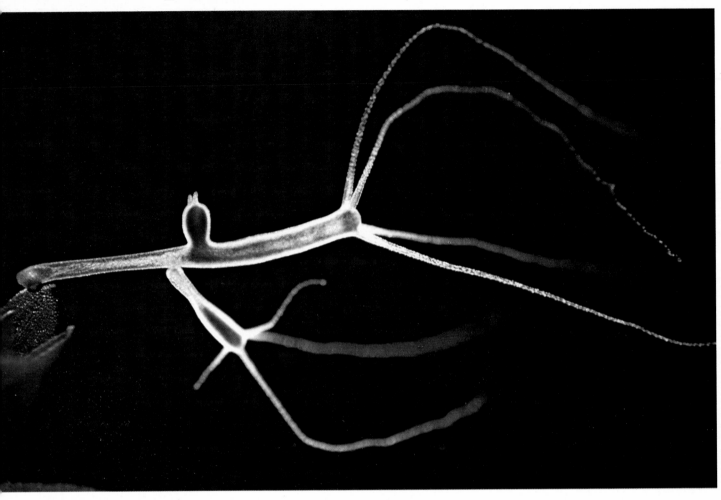

Above: *In the hydra, reproduction occurs when a small swelling, the bud, grows from the parent's body. It develops into a tiny version of its parent and breaks off. Two buds are seen here, one is half developed, the other almost ready to lead an independent life.*

Reproduction is the means of maintaining or increasing the numbers of animals or plants, by generation of new individuals. Its ultimate purpose is the creation of enough of these to spread into new territories and extend the species. Sexual reproduction in animals results in the young animal having a mixture of the characteristics of its parents, and also

perhaps some new features that may help it to be better fitted to its environment.

Asexual reproduction

This method of reproduction may involve part of a parent animal becoming detached and the fragment forming a new individual. This mode of reproduction, known as 'budding', can be seen very clearly in hydra, a small freshwater animal. Another example of asexual reproduction is provided by the planarian flatworm

that can be found under rocks in pools and streams. This animal can actually pinch off the hind region of its body, and the detached piece then quickly develops a head and internal organs, to become a complete flatworm.

Some very simple animals, such as the single-celled animal, *Amoeba*, reproduce by *binary fission* – the act of a cell splitting in half.

It involves two stages: the controlling structure of a cell – the nucleus – divides exactly in half, and then the jelly-like cytoplasm that makes up the rest of the cell follows. The result is two identical cells, each capable of moving away and leading an independent life.

Asexual reproduction is not found in the higher animals, as they are too large and complicated to merely divide in half or grow a 'bud'.

Sexual reproduction

Sexual reproduction involves a cell from the male, the sperm, joining with an egg from the female. From this new joint cell develops the offspring. With some primitive animals it is possible for both sperms and eggs to be produced in one individual. Usually in such cases, however, there is some method of ensuring that they do not meet within the same animal.

Animals show a range of methods of reproduction and care of the young. One of the simplest methods involves shedding eggs and sperm into water and leaving it to chance as to whether they meet. Such animals, which include marine worms, have to produce great numbers of eggs if they are to achieve any success.

In another method the female lays the eggs in a particular place, and the male sheds his sperm over them, thereby improving the chance of fertilization. This is the method used by the stickleback. The male frog actually holds on to the female frog and this means that the eggs and sperm are produced simultaneously so that most eggs are fertilized. Despite this, the frog still makes no attempt to look after the eggs but the eggs are protected in jelly.

Adapting to reproduction on land

When animals first colonized the land they encountered problems with keeping the eggs and sperm moist, and with protecting the eggs from drying up. Some animals such as the frog never solved these problems during the course of their evolution and they still return to water to breed. More successful animals make sure that the sperm do not dry up by placing them within the female's body, where fertilization takes place. The eggs are then either laid after being supplied with a shell, or are kept inside the mother until a later stage of development.

Shelled eggs are laid by such animals as insects, spiders, birds and reptiles. In some cases, such as that of the butterflies, the eggs are not protected and are left to hatch on their own. The sea turtles return to land to lay their eggs, but cannot stay them-

selves, and so they dig a nest in the damp sandy beach and protect the eggs by covering them with a layer of sand. Most

Above: *This female* Euploea cora *can be seen carefully laying its eggs around the young growing tip of the food plant.*

birds not only stay with their eggs to incubate them, but also protect and feed the young after hatching.

You will notice that as the parents take greater care of their eggs and young the number of eggs produced decreases. For example, fishes may lay millions of eggs whilst birds lay very few. Mammalian eggs are fertilized inside the female's body and in most cases the embryos develop there until ready to be born as miniatures of their parents. All mammals feed their young on milk and protect them, and many have long periods of training before the young can live independently.

Left: *Green turtle females* (Chelonia mydas) *crawl up the beach and dig holes in the sand, in which they deposit their eggs before returning to the sea.*

Animal senses

Animals move about for a number of reasons; to feed, to reproduce, or to colonize new areas because of excessive competition. In order to hunt and travel successfully, they need to be fully aware of their surroundings, so that they can behave in a way that is most likely to keep them out of danger.

This information about the world around them comes from their sense organs, such as the ears and eyes.

The extent to which any sense organ is developed depends upon the life-style of the animal, and this in turn is partly controlled by where it lives. Thus, if the differ-

Above: *The clown fish* (Amphiprion frenatus) *produces a slime which protects it from the stings of its associate and host which is a sea anemone. Clown fish gain some protection from predators.*

ent types of habitat are studied, then perhaps we can see just how important a particular sense organ is.

Aquatic animals

Animals living in water, whether fresh water or the sea, all make use of the senses of smell and taste. Small particles of material are carried in the water currents, and these are either smelt or tasted by the aquatic animal.

In the case of animals that do not move, such as sponges, this can be the most important sense as they do not hunt, but merely select what they need from that which drifts by. Other stationary animals such as sea-anemones, not only smell the food, but also use touch to detect their prey. The tentacles of a sea-anemone can feel, smell and in addition paralyse passing prey.

Many of the free-living, active hunters also use their sense of smell. The shark is notorious for its ability to detect blood, for instance. The shark's eyes, like those of most fishes, are of little use in detecting the movements of other animals, and often the water is cloudy and thus light cannot pass through it easily.

A useful means of finding out where a possible predator, or the next meal, might be is the use of pressure detectors. The lateral line of fishes is such an organ and it enables them to detect movements in the water so that they know if something is moving near them. Some of the most efficient forms seem to be able to determine what type of animal is moving, and even how many of them there are!

Balance is important when moving in water, for the weight of the animal is borne by the water, and hence it is easy to drift the wrong way up. Simple balance organs, called statocysts, are found in many animals, such as jellyfishes, lobsters, and even a modified form is found in the ears of more complicated animals such as fishes and whales.

Land animals

The most important sense in air is sight, and the eyes of many animals that live on the land have developed enormously whether the animal is a hunter such as a lion, or the possible prey, such as a deer. Many land animals also retain keen senses of smell and hearing and this helps both hunter and prey.

The keenest sight is found in the birds, and they are capable of seeing for enormous distances and, due to their height when flying, they can scan a large area of land. Their vision assists their ears with balancing and this is very important while in flight.

Bats are also superb fliers. Some operate a sort of sonar, emitting very high pitched supersonic squeaks which bounce off objects in their path and are reflected back to their sensitive ears. This system has developed at the expense of these bats' eyes which are not very efficient.

Other land animals that have poor sight are the burrowing types. In compensation moles have a keen sense of smell, while worms can detect the vibrations passing through the soil.

ANIMAL EYES

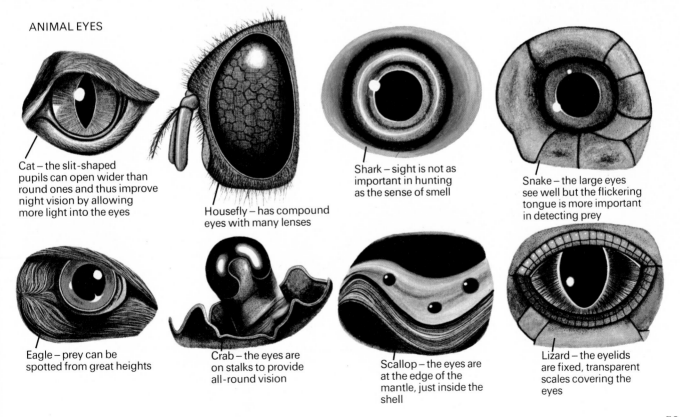

Cat – the slit-shaped pupils can open wider than round ones and thus improve night vision by allowing more light into the eyes

Housefly – has compound eyes with many lenses

Shark – sight is not as important in hunting as the sense of smell

Snake – the large eyes see well but the flickering tongue is more important in detecting prey

Eagle – prey can be spotted from great heights

Crab – the eyes are on stalks to provide all-round vision

Scallop – the eyes are at the edge of the mantle, just inside the shell

Lizard – the eyelids are fixed, transparent scales covering the eyes

Animals in danger

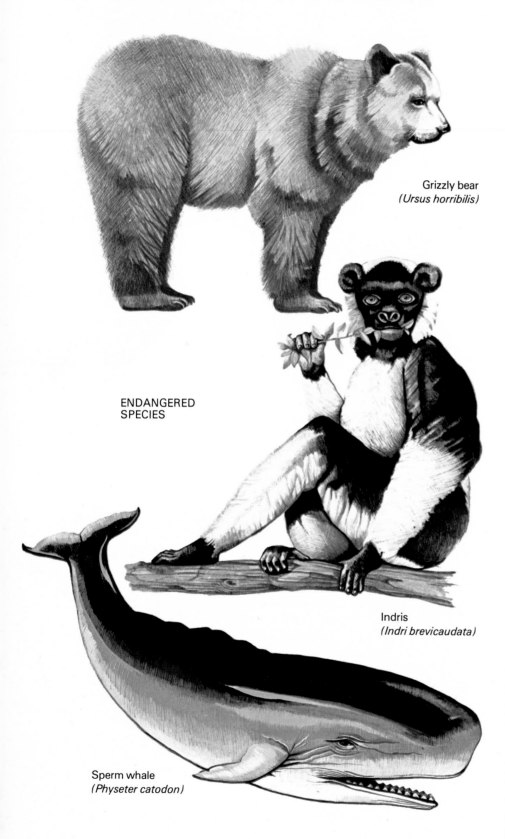

Grizzly bear
(Ursus horribilis)

ENDANGERED
SPECIES

Indris
(Indri brevicaudata)

Sperm whale
(Physeter catodon)

When man was a simple hunter killing only for his food and clothing, he caused little upset to the balance of nature. However, nowadays he exists in vast numbers, often in cities, and thus causes considerable imbalance to his environment.

The amount of land which is covered by buildings is enormous and has required the clearing of thousands of square kilometres of forest. Cities have been linked by roads and railways which have not only taken land, but have divided what was once a single stretch of countryside into smaller units. Such small units are often unsuitable for the larger animals such as the carnivores and hence the numbers of these animals have declined even though man may have had no intention of displacing these species.

The forest clearance has been further extended by the need for farmland, to grow crops and raise livestock. Unbalanced and poor use of grazing land has reduced it to poor quality grassland, and in some areas even to desert-like conditions, following erosion of the topsoil. This is shown in countries bordering the Mediterranean Sea and especially in North Africa where the Sahara Desert has been enlarged due to the grazing of goats, which have denuded shrubs and small trees.

Agricultural land clearance and development has totally changed many countries. For example, in North America the native deciduous forests have very nearly disappeared. The introduction of livestock to an area has upset the natural food cycles, and the carnivores such as wolves and cougars have been ruthlessly hunted. Similarly any animal thought to compete for grazing with man's stock has been attacked. The kangaroo, for instance, has been slaughtered by the thousand because man thought it threatened his sheep's well-being.

Ironically, introduced species have caused more trouble than any native animal, and this is perfectly exemplified in Australia, where introduced rabbits have denuded many grassland ranges. Also, introduced domestic animals, such as cats and dogs, as well as man's vermin, such as rats, have eliminated much of the native Australian marsupial fauna.

Birds of prey have suffered because in modern farming many chemical sprays such as insecticides and fungicides are used. These chemicals collect in the bodies of insects and in turn those of the small carnivores, rodents and small birds for example, that eat them. These small mammals and birds form much of the diet of the large birds of prey and hence the dangerous chemicals find their way into birds such as hawks, buzzards and eagles where they can render the eggs infertile and possibly even kill the bird itself.

The animals that once lived in areas taken over by man have been reduced in numbers either by man killing them or simply because of their inability to adapt to their changing surroundings. Sometimes man killed certain animals because they provided an easy and convenient source of food. This was the case with the North American bison and the dodo of Mauritius. Other animals such as the American passenger pigeon, were killed for the pleasure of shooting and hunting.

Some animals have been, and still are, hunted for their fur or feathers, or some other part of the body. Fortunately today public opinion is rapidly turning against this exploitation of animals.

The growth of industry has increased demands for raw materials, and this has resulted in more destruction of the natural

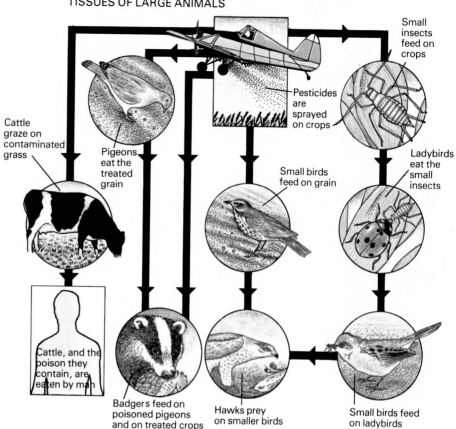

HOW PESTICIDES SPRAYED ON CROPS BUILD UP IN THE TISSUES OF LARGE ANIMALS

Small insects feed on crops

Pesticides are sprayed on crops

Cattle graze on contaminated grass

Pigeons eat the treated grain

Ladybirds eat the small insects

Small birds feed on grain

Cattle, and the poison they contain, are eaten by man

Badgers feed on poisoned pigeons and on treated crops

Hawks prey on smaller birds

Small birds feed on ladybirds

environment and even more species being killed for their products, whales for their oil, for example. The waste products of industry have caused pollution of land, rivers and lakes. The Great Lakes of North America provide a terrible example.

Below: *The demand for whale oil has endangered many species including the sperm whale. Here one is being cut up to obtain the blubber.*

Animal colouration

that was attempting to disguise itself kept moving about and making a noise.

Within one species there are sometimes specimens that are adapted for different backgrounds. For example, the peppered moth (*Biston betularia*) has a light coloured form and a very dark form. The pale moths are found mainly in the country where the houses and trees are

Left: *The milkweed butterfly* (Danas plexippus) *has caterpillars which are distasteful and even lethal to some bird predators. The warning colours have great value therefore to both insect and bird. The poisons come from the plants on which they feed.*

Below: *The caterpillar of the peppered moth* (Biston betularia) *shows camouflage colouration in both texture and colour. At the same time it must behave like a twig, that is stay immobile.*

Animals are coloured for one of three reasons; camouflage, warning, or as part of their courtship display. Camouflage or protective colouration is found throughout the animal kingdom, but is best shown by birds and mammals. It is not only the defenceless prey that is coloured to blend in with the background, but also the hunter. Otherwise trapping and killing the prey would be made even more difficult. Protective patterning is seen very clearly in many deer, and the tiger's stripes provide excellent camouflage when hunting in the shadows of the long grass and bushes of India.

The nestlings of birds that nest on the ground also provide good examples of camouflage. Their behaviour increases the efficiency of their protective colouration: when the mother leaves them they remain absolutely still so that there is no movement for a hunter to detect. It would be foolish if an animal

lean, while the dark ones are found in and around cities where there is more dirt and soot on the buildings and trees.

Courtship is the means by which male and female animals attract each other so that they can come together to mate and produce offspring. Although many animals display very complicated actions during courtship, a lot of them still rely on colours and patterns to render them attractive to a potential partner. For example the peacock has a huge, beautifully patterned tail that he fans out and displays to the female, while the male stickleback becomes very colourful, in the breeding season, with a brilliant red underside. He will attack and chase off any fishes that are the same size and colour, but courts the females, with their yellow undersides.

Deep-sea animals often attract each other by the use of colour, but in this case the colours glow in the dark since there is little or no light in the depths.

Protective colouration

Some animals are able to vary their colouration. Certain arctic species provide good examples of this. The arctic hare and arctic fox, for example, have brown coats in summer and white coats in winter so that they are camouflaged against the different backgrounds of the seasons.

An animal's warning colours serve to indicate to other animals that this is an animal that is either dangerous or unpleasant to eat. Wasps and some snakes have stripes of colour that warn other animals not to touch them. Some harmless animals have copied the colouration of dangerous ones, thereby gaining protection from potential enemies. For example, hover flies are striped like wasps and a lot of people and animals think that they can sting and so keep away from them. In fact hover flies eat nectar and are harmless.

Left: *The vivid patches of colour shown by the peacock butterfly* (Nymphalis io) *serve to intimidate or worry would-be predators. The 'eyes' of the hind wings are most realistic in the warning display.*

ARCTIC HARE

Summer

Winter

Animal classification

In this book the animals are arranged in groups called **phyla**, with simple animals described first and the most highly developed ones last. This is called a 'systematic' classification which shows the evolutionary development of one group from another as the animals of each new phylum are more complex than those of the previous one. Each group is further divided into smaller groups – a phylum is divided into **classes**, a class into **orders**, an order into **families**, and the subdivision of a family is a **genus**. At each stage of classification, the animals within a named subdivision are more closely related and have more characteristics in common. The final level of classification is the **species**. Common names of species vary in different languages eg. English **honey bee**, French **le mellifère**, German **die Biene**, but the Latin name, *Apis mellifera*, is used universally by scientists and identifies the animal exactly.

In earlier sections many anima have been mentioned, most these have been animals th possess backbones. In oth words, vertebrate animals. Vert brates are often large and colou ful and therefore are easily se and familiar. In fact, the vert brates represent only 5 per cent the animal kingdom. There hav been over 1 million different sp cies of animal recorded, and 9 per cent of them do not hav backbones at all. These are th invertebrate animals.

Although invertebrates do n have a backbone they often hav a complex body which is su ported by some sort of skeletor Worms rely on a fluid 'skeletor inside their bodies to keep the in shape; corals use a chalk covering material; snails an clams have shells; whilst th arthropods, the insects, spider and crustaceans, have wha amounts to a suit of armour.

Invertebrates show a wide rang of shape, size, and way of lif Some, like the spiders, snails an earthworms, are very well know to most people. Other types tha are less well known include th parasites, such as the hookworm the leech and the malarial par site. Happily there are also man invertebrates that are looke upon with favour. Oysters an crabs, for example, supply us wit food, and bees not only pollinat flowers but also produce honey.

The first animals to evolve wer simple invertebrates and wer probably single-celled forms There are still single-celled, o unicellular, animals alive, al though these modern types ma be very different from thos earliest ancestors.

As time passed there evolve more complicated animals which were made up of many cells. Wit

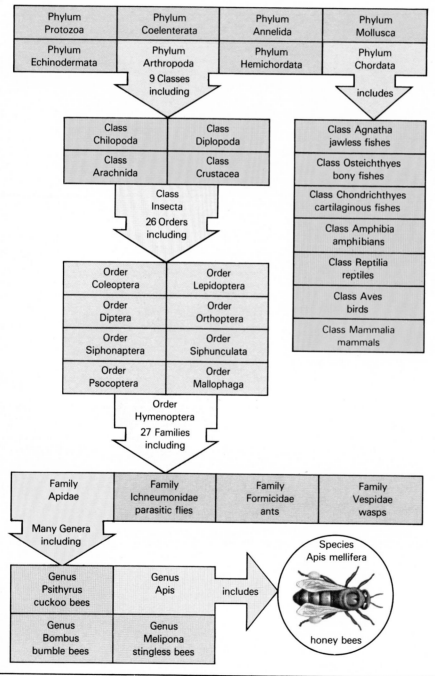

Phylum Protozoa	Phylum Coelenterata	Phylum Annelida	Phylum Mollusca
Phylum Echinodermata	Phylum Arthropoda	Phylum Hemichordata	Phylum Chordata

9 Classes including

includes

Class Chilopoda	Class Diplopoda
Class Arachnida	Class Crustacea
Class Insecta	

26 Orders including

Class Agnatha jawless fishes
Class Osteichthyes bony fishes
Class Chondrichthyes cartilaginous fishes
Class Amphibia amphibians
Class Reptilia reptiles
Class Aves birds
Class Mammalia mammals

Order Coleoptera	Order Lepidoptera
Order Diptera	Order Orthoptera
Order Siphonaptera	Order Siphunculata
Order Psocoptera	Order Mallophaga

Order Hymenoptera

27 Families including

Family Apidae	Family Ichneumonidae parasitic flies	Family Formicidae ants	Family Vespidae wasps

Many Genera including

Genus Psithyrus cuckoo bees	Genus Apis
Genus Bombus bumble bees	Genus Melipona stingless bees

Species
Apis mellifera

includes

honey bees

the development of these multi-cellular forms the invertebrates started to become specialized for different habitats

As different groups evolved many must have failed and died out. Unfortunately, we have only a poor fossil record of these failures since their tissues have not been well preserved in the muds and rocks. During invertebrate evolution there have been a few very important steps which have given rise to the more successful groups.

One very important development was the evolution of the worms, and from them the arthropods. Worms tend to be found only in water or damp soils, but the arthropods have flourished both in water and on dry land.

Another important group is that comprising the molluscs, the snails, oysters and squid for example. This is a very large group and includes the largest of all invertebrates, the giant squid which may grow to 15 metres.

The echinoderms form yet another important group. They include starfishes and sea-urchins which had amongst their ancestors animals which eventually evolved into the vertebrates. This makes the echinoderms of considerable interest to biologists.

Finally it should be remembered that colonization of the land has been most successfully achieved by an invertebrate group of animals, the insects, and not by a vertebrate group. There is no other group of animals anywhere nearly as large in numbers of species as the insect group. Insects represent 75 per cent of all named animals and are to be found in all terrestrial and freshwater habitats.

Right: *This comb from a honeybee hive shows the numerous workers who feed the grub-like larvae visible in some cells in the upper half of the picture. A larva feeds over 1000 pollen meals a day.*

Protozoans (PHYLUM PROTOZOA)

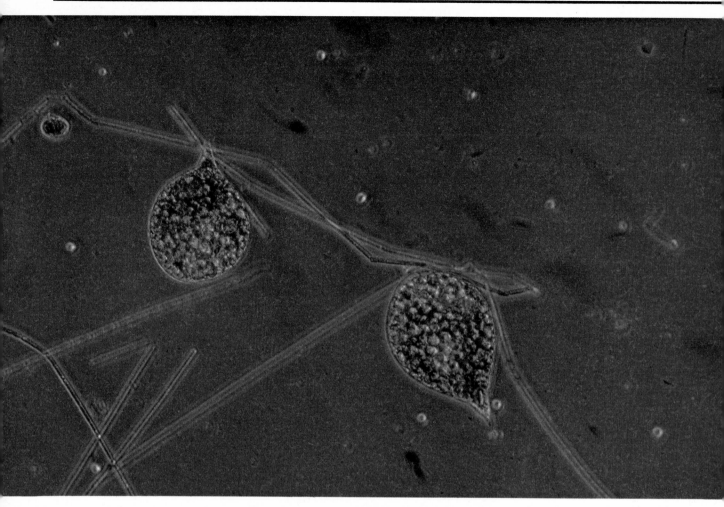

The Protozoa are very small animals which consist of only one cell. This single cell is capable of carrying out all the various functions necessary to sustain life. Some protozoans produce a sort of 'shell' or skin over the surface of the cell. This cellular covering is seen very clearly in such protozoans as the Foraminifera and the Radiolaria. The Foraminifera are economically important as their fossil remains, found deep in the earth, can be used to identify the rock layers and so indicate the possibility of oil deposits.

Some of the simplest protozoans, such as *Amoeba*, show very little structure under a normal microscope, and yet such simple animals are to be found in oceans, ponds and lakes and other species occur even in damp soil. This is a great success by any standards, since it means that this group of simple animals has colonized most of the available sites on the Earth, the only limitation on this colonization being the necessity for water.

Many species can form tough coatings around themselves and thereby exist as spores during periods of drought. They may be blown around in the dust to eventually settle far away from their original home. In this way protozoans can spread rapidly across vast distances.

Not only can protozoans be spread rapidly by wind but they may travel great distances on the

Above: *Many simple organisms are difficult to classify definitely as an anim or a plant.* Astasia *is a protozoan but contains chlorophyll as do plants. It moves by means of its whip-like flagella*

bodies of larger animals. All pr tozoans can increase their num bers very quickly by asexuall reproducing. A single cell divide into 2 and then these divide t give a total of 4 and so on.

Many protozoans feed by e gulfing their food; that is, part their jelly-like body flows aroun the prey trapping it together wit some of the water in which it wa swimming. This drop of liqui containing the prey is called food vacuole, and the food digested within it.

Some useful protozoans a

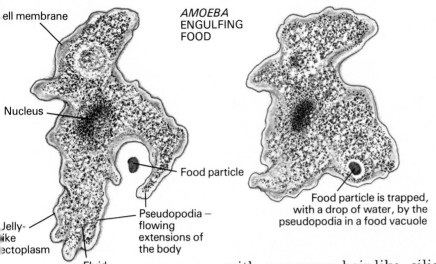

AMOEBA ENGULFING FOOD

ell membrane

Nucleus

Jelly-like ectoplasm

Fluid endoplasm

Pseudopodia – flowing extensions of the body

Food particle

Food particle is trapped, with a drop of water, by the pseudopodia in a food vacuole

wood and plant material for its host. In return *Trichonympha* receives a warm, moist home with a ready supply of food material.

Another useful protozoan is found, probably feeding on bacteria, in the stomach of cattle. The bacteria break down the plant material for the cow, whilst the protozoan provides additional protein for its host.

A close relative of the common freshwater *Amoeba* is to be found as a parasite, living in the intestines of various vertebrate and invertebrate hosts. This organism causes the disease, amoebic dysentery, in man. It can be very serious and may even result in death.

ound in the guts of animals. *Trichonympha* is a protozoan,

with numerous hair-like cilia, which is found in the intestines of wood-boring termites, and it is this protozoan that digests the

---

Parasitic protozoans

One class of protozoans is made up of parasitic forms which cause great harm to man and his livestock: examples of the diseases caused by these organisms are malaria and sleeping sickness. Malaria is caused by the protozoan *Plasmodium*,

which is carried in the saliva of a species of mosquito. When the mosquito bites a human, some of its saliva is injected into the wound, taking with it the *Plasmodium* specimen. This feeds, grows and eventually forms spores. Some of these burst, reinfecting other human cells while others develop into the sexual stage of the parasite. These sexual forms may be

sucked up by a mosquito feeding on an infected patient. The parasite undergoes a series of changes inside the mosquito and eventually the form develops which lives in the salivary glands of the insect and is ready to be injected into another human victim. Malaria patients suffer from recurring fevers and many people die from this disease.

PLASMODIUM LIFE CYCLE

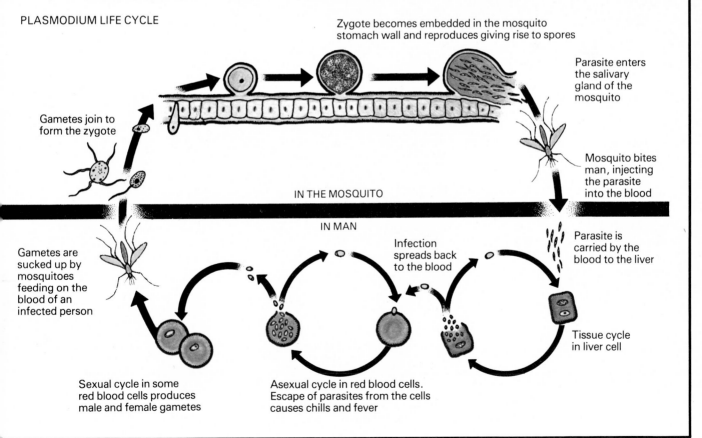

Zygote becomes embedded in the mosquito stomach wall and reproduces giving rise to spores

Parasite enters the salivary gland of the mosquito

Gametes join to form the zygote

IN THE MOSQUITO

Mosquito bites man, injecting the parasite into the blood

IN MAN

Infection spreads back to the blood

Parasite is carried by the blood to the liver

Gametes are sucked up by mosquitoes feeding on the blood of an infected person

Tissue cycle in liver cell

Sexual cycle in some red blood cells produces male and female gametes

Asexual cycle in red blood cells. Escape of parasites from the cells causes chills and fever

Coelenterates (PHYLUM COELENTERATA)

The coelenterates are aquatic animals, and most of them live in the sea. The body of all coelenterates is built from 2 layers of cells with jelly-like material packed in between. In the centre of the animal is a large space which serves as a region for digestion. There is only one opening into the body, the mouth, and this is usually surrounded by tentacles which can often be very long.

Because they are only simple animals and do not have a complicated muscle system the coelenterates are not fast moving. In fact many of them, the corals for instance, spend their lives fixed in one spot. Others, such as jellyfishes, drift passively in the ocean currents and only move actively by pumping water out of folds in their body – a sort of slow and primitive jet propulsion.

Coelenterates may show 2 different stages in their life cycles: the polyp or feeding stage; and the medusa stage, which is the sexually reproductive stage.

Sea-firs and sea-ferns

(CLASS HYDROZOA)

In the Hydrozoa both the typical stages can be seen. The polyps are found in colonies and they feed continuously, producing thousands of tiny *medusae,* which look like miniature jellyfishes. The Portuguese man-o'-war is another colonial form of hydrozoan, and in this case the different members of the colony have become modified in their shape and specialized for one job. For example, some form an entire tentacle, others are shaped to act as 'stomachs

Left: *The Portuguese man-o'-war* (Physalia physalis) *lives in warm tropical waters but occasionally drifts into British waters with the warm Gulf Stream. Although its sting is painful it is rarely fatal to humans.*

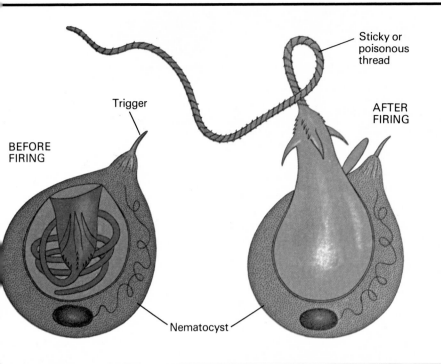

BEFORE FIRING

Trigger

AFTER FIRING

Sticky or poisonous thread

Nematocyst

Coelenterate 'weapons'

Although coelenterates vary a great deal in their shape, they nearly all have stinging cells or nematocysts somewhere in their skin. There are different types of nematocyst: some actually pierce the skin of passing animals and discharge a poison, while others are sticky and merely wrap around the struggling prey. The tentacles usually have thousands of different stinging cells and these help to paralyse and hold the prey whilst the tentacles push it towards, and into, the mouth. Some jellyfishes can cause severe damage to the skin of anyone swimming nearby and touching the tentacles. The Portuguese man o' war (*Physalia physalis*) causes blistering and the sting from some small jellyfishes called sea wasps (*Chironex fleckeri*), found near Japan and the Philippines, has been known to kill adult fishermen within 10 minutes!

whilst others spend their lives in reproduction.

Sea-anemones, sea-fans and corals

(CLASS ANTHOZOA)

The Anthozoa form another large and very important group of coelenterates. Sea-anemones are single large polyps which live attached by their bases to rocks. All anemones have numerous tentacles and they are often brightly coloured.

Sea-fans consist of single polyps which grow very long and thin and bud off secondary polyps along their length. This polyp structure is rather thin and is supported by a chalky skeleton.

The corals form the largest colonial masses in the animal kingdom. Coral reefs, formed from hundreds of different species, can stretch for thousands of

Left: The fleshy body of this sea-anemone polyp is buried at its base by silt but will be firmly attached to the rock beneath. The numerous tentacles show ridges along which the sting cells are situated.

kilometres. They are very important to man, since on the one hand they present a hazard to shipping and on the other they support a rich fish population. The different types of coral formation were mentioned in the section on the seas.

Jellyfish

(CLASS SCYPHOZOA)

The familiar jellyfishes consti-

Above: *This soft coral is composed of many small polyps linked by internal canals. They get their name because their skeletons are of tough gelatinous material, unlike the calcareous reef-forming corals.*

tute a third group, the Scyphozoa. They spend most of their lives as medusae drifting in the oceans, moved by the currents and winds. Jellyfishes range in size from a few millimetres to just over 2 metres across. The giant forms found in the north Atlantic may have tentacles up to 60 metres long!

Molluscs

The molluscs form the second largest group in the animal kingdom; only the arthropod group has more species. Molluscs are extremely variable in size and shape and hence have only a few features in common but many of them possess a shell which acts as a protective skeleton.

The shells may be round and smooth like those of the common snails, coiled into a spire, curved like a tooth, fan-shaped or even covered in spines. In some cases, the shell is made from a pair of flaps hinged together, as in the clams and oysters. In other ex-amples, such as the cuttlefish (*Sepia*), the shell has become internal and is found in the middle of the body.

Inside the shell, molluscs are divided into two main parts, the body containing the gut and re-productive organs, and the mus-cular foot. The foot is used for a variety of jobs. Snails use it for creeping along the ground and vegetation; the squid and octopus have a modified one for catching prey, whilst many clams use theirs to dig the holes in which they live.

There are several classes of

Above: *This cuttlefish has just caught a shrimp, the feelers of which are still visible. The cuttlefish have internal skeletons of a chalky material to which the muscles are attached. The siphon propelling the cuttlefish is below the eye.*

mollusc, the smallest and sim plest of which is composed of th 'chain mail' molluscs or chiton (CLASS POLYPLACOPHORA). Thes simple forms are found at the edg of the sea, usually clinging to fla rocks, protected by their har armour of 8 plates of shell

Equally odd is the secon group, composed of the elephan tusk shells, or scaphopods (CLAS

in an unpleasant way, is the damage that they cause to mud embankments with their burrows and the enormous ravages that the shipworm, a bivalve called *Teredo*, creates in wooden jetties, boats and dyke gates. *Teredo* settles into the softer outer layers of wood when still a young larva, and then it spends the rest of its life drilling holes through the timber, using the scraps of wood as its food. The drilling action is achieved by regular rubbing movements of the 2 valves against the wood, the valves having saw-like teeth on their leading edges.

The mussel is a mollusc familiar throughout the world and it plays an important part in the diet of many people living by the sea. The huge mussel beds form when the young mussels attach themselves to the rocks using thin, but very tough, threads. These threads, in addition to the mussels themselves, help to trap fine sand and silt, causing large banks to build up. On parts of the east coast of the United States there are vast reef-like structures, built by generations of mussels which settle on the empty, but still attached, shells of their ancestors.

SCAPHOPODA). These animals are exclusively marine and live buried in the sand into which they burrow using the cone-shaped foot.

The third group of molluscs contains the bivalves and lamellibranchs (CLASS BIVALVIA). These get their name from the shell, which is divided into flaps, or valves, hinged along their back. There are large muscles to open and shut valves, and the foot

Above: *Mussels feed by drawing in a current of water from which they sieve off any fine particles of organic rubbish. It is possible to see the extended tentacle-like structures which help to 'taste' and check the incoming flow of water.*

can project between them in order to dig and feed.

This group is very important economically as a source of pearls, food and decorative shells. Equally important, but

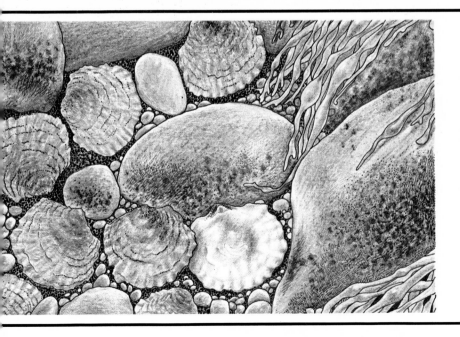

Oysters and pearls

Oysters are eaten by many people, but their fame lies chiefly in their ability to produce the precious pearl. Pearls are produced when some piece of foreign material, such as a fine sand grain, irritates the delicate layer of flesh that produces the shell. The irritating particle is surrounded by layer upon layer of the characteristic lustrous material.

Today many of the natural oyster beds have been exhausted and oysters are raised in special areas around the coasts. In Japan, pearls are cultured by artificially introducing fine grit into young oysters (*Pinckada* species) which are placed in wire baskets and then left for 2 years, after which period of time it is hoped that a valuable pearl may be removed.

Many-armed animals

The last class of mollusc is the cephalopods (CLASS CEPHALOPODA), which contains some of the most complex of all invertebrates. The octopus, squid (illustra-ted), cuttlefish and *Nautilus* are cephalopods. The giant squid (*Architeuthis*) has a body which may grow up to 6 metres in length and, with the tentacles it may reach an overall length of 16 metres. It is the largest invertebrate. All squids are voracious hunters of fishes, whilst the giant squid itself falls prey to the sperm whale.

Octopuses are slower-moving cephalopods, living typically in crevices on the bottom. Like the squid and cuttlefish octopuses have suckers on their tentacles but unlike the other forms, the octopuses have only 8 arms.

Left: *This squid* (Meleagroteuthis) *has 2 sorts of tentacles – the 2 long thin ones which shoot out to sieze prey and the 8 thicker muscular arms like the arms of the octopuses.*

Other important bivalves include the cockle and the scallop. Both these molluscs are eaten by man and can be found in most shops selling shellfish. The scallop is one of the most active swimmers of British lamellibranchs; it moves by clapping its valves together. To help it in its movement there is a fringe of sensory tentacles around the edge of the shell and a large number of very obvious blue-green 'eyes' (see page 53).

The largest of the 7 classes of mollusc is the gastropods (CLASS GASTROPODA). Gastropods typically have a single shell that is coiled to some extent, and out of which extends the large muscular foot for locomotion, and a head that has pairs of tentacles and a pair of eyes. The common garden snail (*Helix pomatia*) and the seaside whelk (*Buccinum undatum*) are both good examples.

Most gastropods live in the sea or along the edges of the sea. Such common animals as winkles and whelks are gastropods that are useful to man as a source of food and that have an important place in the food web. Winkles graze on the algae whilst whelks are carnivores that will eat live prey as well as scavenging for carrion.

The ormer, or abalone (*Haliotis* species) as it is known in North America, is sought after, partly for its food value, but also for its attractive shell. This can be used as an ornament or can be cut up into decorative artwork for inlays, buttons and buckles, for example.

Some of the largest gastropods are the tritons, which are found in tropical seas. Their shells can be as large as 45 centimetres and have been used as war horns. Conches are the largest molluscs

around North American shores, being up to 25 centimetres in length. Again the shells have been used as horns, and today are still used for decorative purposes.

A group of gastropods has developed a primitive respiratory organ for living on land, and this is the group in which are found the very common and damaging

Above: *The extraordinarily beautiful colours of the sea slugs are exhibited by this tropical form as it crawls over some dead coral. The coloured flaps of flesh increase the body surface for respiration.*

garden snails and slugs. It is not only the garden slug that has lost its shell; some of the most beautiful of all marine animals are the shell-less sea slugs.

Leaping cockles

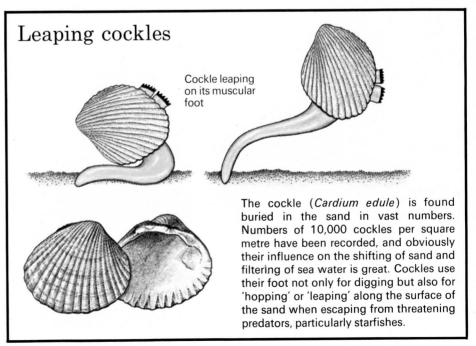

Cockle leaping on its muscular foot

The cockle (*Cardium edule*) is found buried in the sand in vast numbers. Numbers of 10,000 cockles per square metre have been recorded, and obviously their influence on the shifting of sand and filtering of sea water is great. Cockles use their foot not only for digging but also for 'hopping' or 'leaping' along the surface of the sand when escaping from threatening predators, particularly starfishes.

67

Segmented worms (PHYLUM ANNELIDA)

The annelids are the true worms and all of them have their bodies divided into short lengths, or segments. These play an important part in the worm's movement. By contracting the muscles in a few segments at a time, the worm can wriggle and move its body along.

All worms are really only suited to living in water, or at least very damp regions. This is because their skin is very thin and the worms breathe through it. It is partly because worms are so defenceless that they play an important role in the food cycle – providing a rich supply of protein for such predators as fishes and birds. To overcome the dangers of being eaten, many worms have adapted to life in tubes, either

burrows in mud or special structures built from slime and sand. The earthworms, of course, are well known for their burrows.

Bristleworms

(CLASS POLYCHAETA)

The largest group of annelids is that composed of bristleworms, which get their name from the small stiff bristles, or *chaetae,* that are found on their bodies. Usually the bristles are attached to paddle-like flaps which occur in pairs along the length of the body. The paddles are used when the bristleworms swim, often just above the seabed, for this group of annelids are all marine. The brist-

les help the worms to dig into sand and mud, and also serve as anchors. In the case of a West Indian bristleworm (*Chloeia viridis*) the animal can inflict severe stings with the clumps of stiff needle-like bristles.

As mentioned earlier, some bristleworms live in burrows, some form tubes and others swim freely, although these latter worms usually crawl into cracks in rocks for shelter. The tropical palolo worm (*Eunice viridis*) is a good example of a bristleworm that can swim freely, but in fact spends most of its life in crevices in a coral reef. It retreats into these whenever fishes approach. The palolo worms are well known because, at certain times of the year, their tails, filled with eggs

or sperm, break off and rise to the surface to mate. The swarming is so regular, being governed by the phase of the moon, that natives can predict the appearance of these worms and paddle out in canoes to collect them as food.

The common ragworm (*Nereis diversicolor*), shown right, may be 10 centimetres long and is one of the commonest bristleworms to be found around the shores of Europe. It lives in rock pools, at the tide's edge, and may be discovered buried in silt or swimming across the bottom. It has powerful jaws that are capable of biting large pieces of flesh from any dead animals that it finds. The picture shows the numerous flaps and tentacles that help the ragworm find its food.

The lugworm (*Arenicola marina*), another very common worm around European shores, is one of the many bristleworms that live in burrows. In this case the burrow is lined with slime to prevent it collapsing. The lugworm uses its paddles to draw

Above: *The common ragworm is well known to fishermen as a bait. Again the many paddle-like gills and bristles can be seen. The head (top right) has many sensory probes and a pair of eyes on fleshy lobes.*

water in at one end of the burrow, the water is then filtered for food, and then expelled at the other end, by the action of the paddles. This system of feeding is very common amongst a number of tube-dwellers.

Earthworms

(CLASS CLITELLATA)

Earthworms have adapted to a life of burrowing in the soil by losing the eyes, tentacles, and any projecting flaps that might slow down their burrowing. They are of importance because their burrowing mixes, aerates and drains the soil, and they serve as a food source for birds, moles and other animals.

Leeches

(SUB-CLASS HIRUDINEA)

The leeches live mainly in fresh water and warm wet areas and they have become specialized for feeding on the blood of other animals, which may range from pond snails to large mammals.

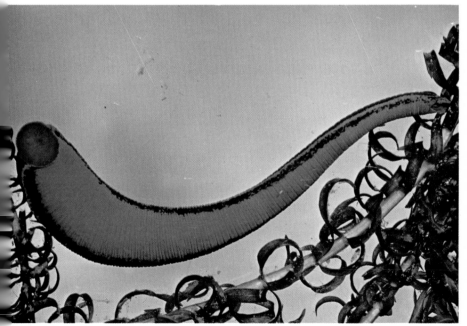

Left: *The medicinal leech* (Hirudo medicinalis) *is found throughout Europe, although only in the Lake District and the New Forest in Britain. The mouth lies in the middle of the front sucker and an active feeder may take up to 5 cubic centimetres of blood.*

Onychophora (PHYLUM ONYCHOPHORA)

This group forms a living link between the annelids and the arthropods. Because they have some annelid-like features, they are not placed within the Arthropoda but form a separate group. The commonest members of the Onychophora are the various species of *Peripatus*. They are all tropical animals found mainly in leaf litter, under the bark of rotten trees or even in damp soil. Certainly the habitat must be warm and moist. They are nocturnal and eat small insects, woodlice and any dead animal remains that they may find.

Right: Peripatus *is seen here scuttling for cover as it seldom comes into the open for long. It detects its prey using sensory antennae and on finding food ejects slime from glands on its head to trap prey.*

Arthropods (PHYLUM ARTHROPODA)

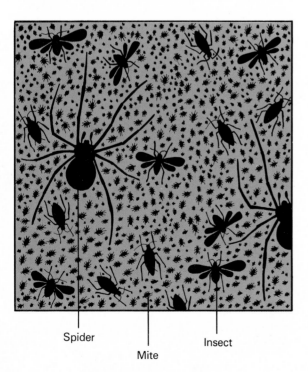

Spider

Mite

Insect

Left: *In a piece of meadow measuring 7·5 centimetres square, you could expect to find 1·43 spiders, 14·3 insects and more than 858 mites.*

This is the largest group of animals in the world, containing about 1 million recorded species, and probably many tens of thousands more that are yet to be identified. Arthropods have colonized every possible habitat on this planet, and they are remarkable not only for the number of species but also for the tremendous number of individuals. For example an acre of meadow land may well have over 1 million spiders, 10 million insects and over 600 million mites!

The arthropods were one of the first groups of animals to successfully move on to land, helped, as mentioned earlier, by their strong tubular and waterproof exoskeletons. In addition to moving with-

Animals with external skeletons

In a group having a million different types of animal it is difficult to find many features that are common to all of them. However, one of the most obvious is the type of limb. All arthropods have jointed limbs that are divided into short lengths with a hinge between each section. The sections of limb, and indeed the entire body, are usually covered in some form of protective shell or exoskeleton. This exoskeleton is the feature that is responsible for the great success of the arthropods.

The arthropods' ancestors were primitive segmented worms that still have descendants alive today in the large group of marine bristleworms. The segmented body of the ancient worms is still seen as the third feature common to all arthropods.

The joints of the legs and the hinges between the body segments are needed because of the stiff nature of the exoskeleton. It cannot bend and therefore, if the animal is to be capable of moving on its legs or bending its body or jaws, there must be hinges. The muscles moving the various parts of the body are protected inside the tough exoskeleton.

In some arthropods where weight is not important, for example the aquatic crustaceans (CLASS CRUSTACEA), the already strong exoskeleton is further strengthened with heavy deposits of chalky grains. Thus it may be very dense, but, as the animal is supported by the water in which it lives, the weight does not restrict its movement.

Other, land-living arthropods cannot support such a heavy exoskeleton and have developed instead a thinner but extremely tough shell with a layer of waxy material on the outside. This greasy layer prevents the owner drying out in the sun, and this feature has permitted arthropods such as insects, spiders and scorpions to colonize the very dry areas of the world.

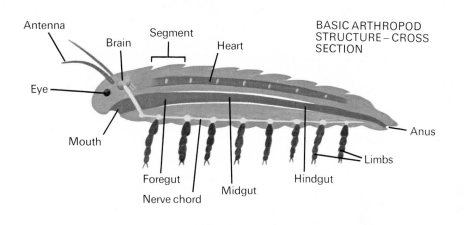

BASIC ARTHROPOD STRUCTURE – CROSS SECTION

Antenna · Brain · Segment · Heart · Eye · Mouth · Foregut · Nerve chord · Midgut · Hindgut · Limbs · Anus

out drying out, it is necessary for a land animal to breathe easily through respiratory organs that are adapted to breathing the oxygen in the air. The aquatic animals, of course, breathe either through their skin or with gills. Terrestrial arthropods use a range of respiratory methods. The insects have a series of internal air tubes, or tracheoles, while the spiders have a series of flat plates called lung books.

The classification of the Arthropoda is very complicated due to the huge numbers and the range of shape and size. However, there are 6 main groups generally recognized within the Arthropoda.

Trilobites

(SUB-PHYLUM TRILOBITOMORPHA)

The members of this group of marine arthropods are all extinct: they lived between 350 and 500 million years ago. During their time they were a dominant and important part of the fauna that crawled and swam through the ancient seas. They showed the typical arthropod segmentation of the body, it being divided into an obvious head with the thorax and abdomen looking very similar. In some cases, the body could be rolled up, and this was most probably a defence mechanism that we still see in many animals of today, the woodlouse, hedgehog and pangolin, for example.

Below: *This fossil trilobite shows the characteristic segmentation of the abdomen. It is thought that this type of arthropod pushed its way through and over the surface of marine muds feeding on scraps.*

Arachnids

(CLASS ARACHNIDA)

The arachnids often cause people to shudder and express horror, and certainly few classes of animal have ever had a worse public image than this group, made up of spiders, scorpions, mites and ticks.

Fossil scorpions (ORDER SCORPIONONES) have been found in rocks over 400 million years old, and this makes them the most ancient of the land-living arthropods. They possess a powerful pair of pincers and the notorious sting on the tail. Few species are actually dangerous to man, although several can inflict a painful wound. Some species are successful colonizers of deserts, whilst others are found in humid rainforests. Obviously this is a group that shows a variety of adaptations to different ways of life, although they all require warmth in which to live.

The spiders are an immensely

important group of carnivores. They do man a lot of good by eating grubs and insects that would otherwise do him a great deal of harm. Many species produce silk from special glands and with it weave webs in which prey becomes caught, to be killed by the venomous bite of the spider. Others do not weave webs but live in short burrows, some of which even have carefully fitted flaps. These are the trapdoor spiders. The real trapdoor spiders are tropical animals, but there are some close relatives in the temperate zone called purse-web spiders.

Mites and ticks (ORDER ACARI) are the most important of all arachnids, due to the parasitic habits of many of them. Not only do they do damage by removing blood from many hosts, including man, but they also carry organisms that cause disease. Hence the bite of a tick not only means irritation and loss of blood, but also possibly sickness. Scrub typhus and spotted fever are spread in this way, for example. The red spider mite (*Tetranychus* species) causes great damage to

Above: *The imperial scorpion displays th sting in its tail segment very clearly. The pairs of walking legs help it to scuttle int rock crevices whilst the fearsome pincers seize any small animals.*

fruit trees and greenhouse crops

Mites are the cause of muc discomfort to man and they ca cause very unpleasant allergie in some people. Their dead bodie and exoskeletons in furniture an mattresses form part of the dust i the house, and this causes som people to cough and even result in rashes in people with ver sensitive skins.

Other arachnids include th harvestmen (ORDER OPILIONES) They are not true spiders, al though they closely resembl them. They kill their prey, no with venom like true spiders, bu with powerful jaws. Many are om nivorous and hence eat a rang of plant and animal remains.

Centipedes and millipedes

The centipedes and millipede were once included in the sam class of Arthropoda, but they ar now known to be very different i many ways. The millipedes ar herbivorous, living on dead plant

Left: *The red spider mite thrives in hot dry conditions and causes damage to greenhouse crops and fruit trees by sucking juices from the underside of the leaves and soft stems.*

remains with an occasional meal of young seedlings. A few species are known to eat rotting animal tissue, and hence act as scavengers.

All millipedes (CLASS DIPLOPODA) have tubular bodies with very obvious segments, each of which appears to have 2 pairs of legs. Despite their name and the commonly held belief, they do not have 1000 legs – the maximum number being 200. They move with waves of activity passing along the rows of legs as they push their way through leaf litter and behind bark.

Millipedes are known from all over the world, wherever there is a supply of damp plant remains. Although the temperate species are seldom larger than 3 centimetres, the tropical ones can grow to 30 centimetres. Such large, slow, and apparently helpless animals defend themselves with rows of glands along their sides which produce a foul smell to deter their attackers. All have tough exoskeletons and some, the pill millipedes, can coil up into a ball, presenting only a row of curving, heavily armoured plates to a predator.

Centipedes (CLASS CHILOPODA) are fast-moving hunters of the soil and leaf litter. They look like millipedes, except that their bodies are flattened and they only have 1 pair of legs, albeit long ones, per segment of the body. They have powerful jaws which inject a poison into their prey to kill it. Small insects, earthworms and any grubs present among the decaying leaves and twigs are eaten by these mainly nocturnal hunters. The largest species grow to just over 25 centimetres in length and can kill small mammals, nestling birds and lizards, as well as worms and insects.

Below: *This huge centipede from Botswana has massive poison fangs which can be seen as brown structures bulging at the sides of its armoured head.*

Fearsome scorpions

The solpugida, the sun spiders or wind scorpions, form a more vicious group of arachnids. These fearsome looking animals, with their huge jaws, live in hot regions, especially deserts in Arabia, Africa and parts of North and South America. They are insatiable, savage carnivores and, although small (between 1 and 5 centimetres long), they will attack and eat small rodents, birds, lizards and other arthropods. They use their huge jaws to kill prey, but, although they can give a very nasty bite, they are not dangerous.

Left: *The enormous jaws of the sun spiders can easily cut through a large tropical locust. The jaws are unusual in being paired and biting vertically. The swollen bases of the jaws contain muscles. Sun spiders are not poisonous but instead rely on stealth, speed and above all their terrible bite to obtain their meals.*

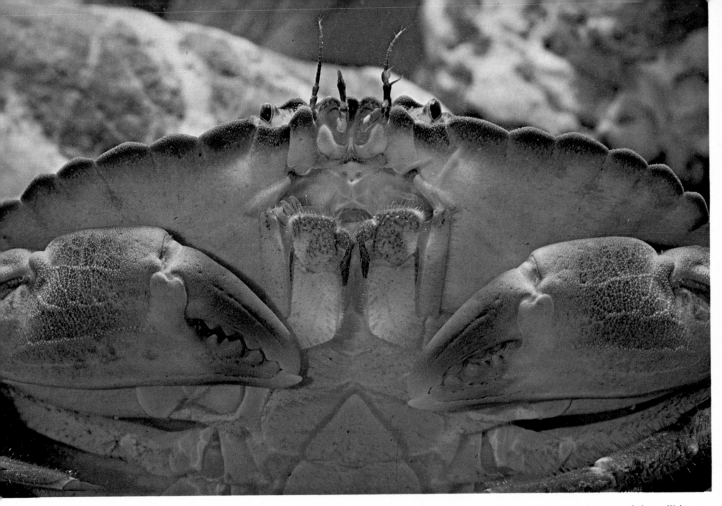

Crustaceans

(CLASS CRUSTACEA)

These form a very large group of arthropods, having over 26,000 species distributed throughout the world. They are mainly aquatic with many marine and freshwater forms. There are not many terrestrial species and, although a land-dwelling way of life may yet evolve, these crustaceans would have to compete with the insects, which on the whole are an established terrestrial group.

A feature of the crustaceans is the specialization of the legs. Each body segment has a pair of limbs, which are particularly well developed in the head region, and are adapted to a variety of functions. There are 2 pairs of antennae which are very obvious and are used to probe their surroundings. The eyes are usually on stalks and may be swivelled to

follow changes in light. The mouthparts vary greatly: the crabs have large pincers for holding and tearing prey; many shrimps have comb-like limbs for sieving out food, and many parasitic forms have needle-like processes for piercing an unfortunate host.

Another feature often seen in the large crustaceans is the massive shell which covers the head and thorax. This is very heavy, and protective in function, and is missing in simpler forms.

Some of the smallest crustaceans are the copepods and water-fleas. The water-fleas (*Daphnia*) are well known to people who keep aquarium fishes because of the widespread use of *Daphnia*, the common freshwater species, as fish food. The different species of water-flea occur in huge numbers in ponds and lakes and provide an enormous mass of food for insects, fishes, birds, worms and any number of other

Above: *The mouth parts of the edible crab* (Cancer pagurus). *The large pincers for holding and tearing food help to protect the underside of the head. The 2 pairs of short antennae help the eyes in detecting food.*

water-dwellers. The copepods are equally important in the life of the oceans since a very large proportion of the zooplankton comes from this group. These copepods provide food for such important commercial fishes as the herring, as well as for the persecuted whalebone whales.

In addition to providing an essential food source for larger animals, the copepods have relatives which are far from helpful to fishes. There are many parasitic forms which live in and on their gills.

Other groups of crustaceans also have unpleasant parasitic members, including some which are parasites of other crustaceans. The barnacles are harmless crustaceans which lead their ntire adult life attached to rocks, jetties and ships' timbers. Although they make life painful for holiday makers, they do not hurt other marine animals and they feed by filtering out fine food particles from the water. However, there is a relative of the barnacles, called *Sacculina* that invades young crabs, especially the common European shore crab (*Carcinus maenas*). It grows in the abdomen of the unfortunate host, and eventually eats away most of the crab, which naturally dies, but not before *Sacculina* has grown so large that it bursts out through the body of the host.

The best known of all crustaceans are the shrimps, crabs and lobsters seen so commonly around the shores of the world. In fact, species have been recovered from all the various depths of the oceans. Some of the deep-sea shrimps possess patches of skin that glow in the cold, dark waters, possibly helping the shrimps to stay together. Others, such as the spider crabs, are common dwellers on the continental shelves. One species of Japanese spider crab grows to reach a body diameter of over 15 centimetres with a leg span of just over 2.5 metres! The tropical shores teem with different species. The unmistakeable fiddler crabs (*Uca annulipes*) are found in mangrove swamps and muddy shores, the male busily waving his outsize claw to attract the female and defend his territory. The land crabs of the tropics, as their name implies, live much of their lives on land, where they eat insects and grubs. Their young spend some of their life in water.

Hermit crabs do not have a complete tough exoskeleton over

their abdomen and hence they protect this region by living in abandoned shells of winkles and whelks. An interesting type of hermit crab is the robber crab (*Brigus latio*) which comes on to the land and climbs trees in the mangrove swamps. The adults give up wearing their snail shells and the skin toughens. These forms are so completely adapted to land life that they drown if held underwater for too long.

Woodlice are the best-known terrestrial crustaceans and they occur throughout the world, living on young plants, leaf litter and dead wood. They have lost their gills, their respiration having become adapted to cope with air, although they still live mainly in damp regions.

The final, and by far the largest class of the arthropoda, is the Insecta, and is considered in the next section.

Below: *The male fiddler crab rises on to the tips of its legs in order to gain further height for its famed 'snapping' display using an enlarged left pincer. The eyes peer about on the ends of long, thin stalks.*

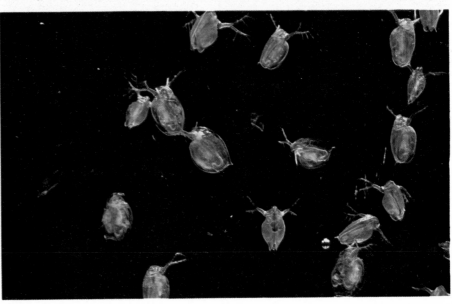

Right: *Daphnia, the water-flea, is a common member of the freshwater zooplankton and forms an important link in the food web of freshwater ponds. It feeds by sieving out fine particles of food and itself is swallowed by fish.*

75

Insects

(CLASS INSECTA)

Insects make up the largest class of arthropods, and there are more species of insect than all other animals added together.

The study of insects is called entomology, and it is extremely important since these animals do an enormous amount of damage to man, his animals and crops, his stores of food and materials, and even his houses and factories. However, despite the terrible damage and disease that is caused by insects, there are some things that they do that are of benefit to mankind. Firstly, insects play a very important role in the pollination of flowers and crops. Also some insects produce commercially important products: for example honeybees produce large amounts of honey and wax, whilst silkworms (*Bombyx mori*) produce the fine silk threads that are used in the manufacture of expensive fabrics.

Before listing the different types of insect and some of the important features of their lives, it is as well to look at features that are to be found in the majority of insects. As they are arthropods, insects naturally have obvious exoskeletons, jointed limbs and segmented bodies. In addition, their bodies are divided into three distinct regions – the head, the thorax and the abdomen.

The head usually bears a pair of jointed feelers, or antennae, which are used to locate food and also to warn the animal of approaching danger. At the base of the antennae are a pair of eyes, each of which is made up of hundreds of tiny lenses, which means that the image of whatever the insect looks at is composed of many little dots in much the same way that a picture is made up of dots in a newspaper. Eyes of this kind, with many small lenses, are called compound eyes. The mouth parts, or jaws, vary enormously and are modified to be best suited to the special diet of each particular insect, as will be seen later.

The second region of the body is the thorax and this consists of 3 segments each of which bears a pair of legs. Thus insects, with 6 legs, can easily be distinguished from spiders which have 8. In addition to the legs, the second and third segments of the thorax each have a pair of wings. In the

Below: *The numerous six-sided lenses of this fly's eyes show up very clearly here. The large size and curvature of the eyes ensures very good all-round vision. The fine hair-like structures also help in touch and smell sensitivity.*

case of some insects, these wings are covered in scales which may give them a range of beautiful colours, whilst other forms have only very narrow transparent wings. Again, these different sorts and many others will be described later.

The segments of the abdomen have no legs but contain the various organs of the reproductive system and much of the gut. In some insects there is some structure at the hind end of the abdomen. This may be for egg laying, as is seen for example in the bush crickets, or for inflicting a painful sting, as is the case with wasps and the hornet (*Vespa crabro*), for instance.

Right: *The painted lady butterfly* (Vanessa cardui) *is well known in Europe, both for its beauty and its long migrations to North Africa. The long antennae with clubbed ends are typical of most butterflies.*

Insect mouthparts

All insects are equipped with the same basic structures which form their jaws and various other parts that are used in feeding. However, during the hundreds of millions of years over which they have evolved, insects have taken to eating a wide variety of different foods, and this has resulted in the mouth-parts, or jaws, developing in

different ways, depending on the particular diet. Most insects eat in one or two ways, either they bite off pieces of food and chew them up, or they suck up liquids such as nectar or blood. The cockroaches have large mouthparts, with a pair of powerful jaws, or mandibles, each of which has sharp pointed cutting edges that are like the blades of a saw. This type of cutting and chewing jaw is found in other types of insect such as wasps, locusts and many scavenging beetles.

Mosquitoes are notorious for their unpleasant and irritating 'bites' which are

inflicted by specially pointed tube-like jaws. There are many other insects which have this type of piercing and sucking jaws, for example the tsetse flies, lice and fleas which are animal parasites, and the aphids and thrips which attack plants.

The housefly (*Musca domestica*) has a strange sponge-like pad on the end of a stalk. This stalk is lowered on to its food and the pad then has saliva passed on to it from tubes which run down the stalk. The saliva softens the food and the semi-digested fluid is sucked back up and into the gut.

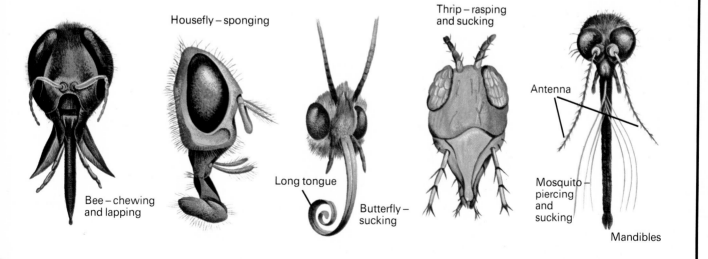

Housefly – sponging

Thrip – rasping and sucking

Antenna

Bee – chewing and lapping

Long tongue

Butterfly – sucking

Mosquito – piercing and sucking

Mandibles

A highly specialized hunter

One of the most highly specialized methods of hunting is found in the praying mantis (*Mantis religiosa*) which sits motionless until its prey comes within reach. The mantis, camouflaged by its green colouration against the leaves, sits on a twig, holding on with the second and third pairs of legs. The first pair of legs is folded up in front of its head. These legs whip out like a pair of living traps to grasp the unfortunate victim, which is then held tightly and is partly impaled on the savage spikes and hooks on the 'forearm' regions of the mantis.

Insect legs

The jointed limbs of insects have also evolved into many different forms. The cause of this diversity is partly the different feeding habits, and partly the particular method of locomotion. The simplest type of legs can be seen in many of the beetles and cockroaches, where all 3 pairs are very similar in structure and are used solely for walking or running. However, many other insects have the last pair of legs modified so that they can be used for jumping. The grasshoppers, locusts and crickets show this adaptation very clearly. The flea is also well known for its power of jumping and has even appeared as a source of entertainment in the 'flea circuses'.

The legs of water beetles and bugs are usually flattened, and they often have a fringe of hairs to make them wider. They then act as paddles as the insects swim through, or skim over, the waters of a pond.

Mole crickets are magnificen burrowers, and like the mamma from which they take their name their front legs are broad and very powerful. They are equipped with sharp edges with which they can cut through the soil and any thin roots that impede their tun nelling.

Wings of insects

The wings of insects are pieces of the exoskeleton that are pulled

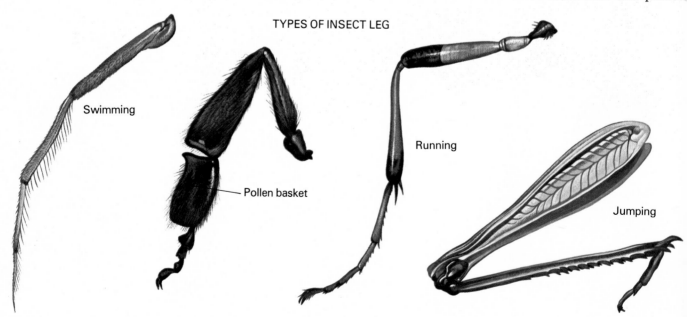

TYPES OF INSECT LEG

Swimming

Pollen basket

Running

Jumping

out into very thin, flat sheets and are strengthened by fine tubes or veins. They are therefore very different from the wings of the other two groups of animals that exhibit true flight – the bats and birds. In the case of the birds, it is the entire front limb that is used, whilst in that of the bats, the fingers are extended with a leathery skin joining the thin finger bones.

The scales on a butterfly wing

The insect wing has thus not developed by replacing other limbs, but merely by extending part of the shell. Most flying insects have 2 pairs of wings, and in the beetles the first pair serve a protective function in that they cover the thin, delicate second pair that are unfolded only for flight.

The true flies have lost their second pair of wings: they have become modified to form balancing organs. Thus only the front pair are used for flying.

Butterflies and moths have wings that are covered in scales which overlap one another in a similar fashion to roof tiles. When touched, these scales are very easily rubbed off.

Insect life histories

One of the reasons for the success of the insects lies in their great range of life history. Most insects lay eggs, but what emerges from the shell can vary enormously.

The most primitive forms hatch as miniature adults and grow up through a series of moults, or *ecdyses,* getting larger after each shedding of the old 'skin'. A common example of this very simple type of life history is provided by the domestic silverfish (*Lepisma saccharina*). This common little insect is found in bathrooms, amongst old books and papers, or anywhere else in a house where it is dark, reasonably damp and there are crevices and a source of starchy food. The little silverfish may moult up to 50 times in the course of its life.

The two other sorts of life history are more complicated and the young that hatch are unlike their parents.

In one type, the emerging young are called nymphs and these have a body shaped roughly like that of the adult. However, they lack wings, and internally they have not matured as they have no reproductive organs.

Below: *Demoiselle, or damsel, flies are related to dragonflies. These, of the genus* Agrion, *are frequently found swooping over vegetation at the sides of streams where they catch their prey. They are good flyers and their large eyes help in hunting.*

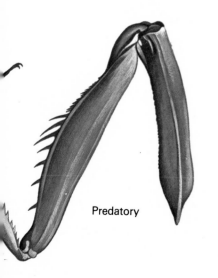
Predatory

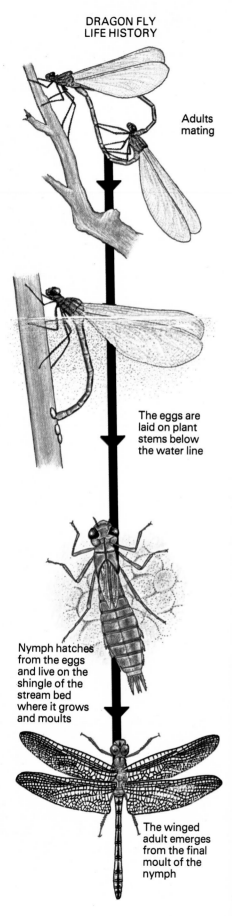

DRAGON FLY LIFE HISTORY

Adults mating

The eggs are laid on plant stems below the water line

Nymph hatches from the eggs and live on the shingle of the stream bed where it grows and moults

The winged adult emerges from the final moult of the nymph

Nymphs eat voraciously and are some of the most savage insect carnivores. The dragonfly nymph will eat virtually anything that it can catch and hold in its jaws. It consumes insects, as well as small vertebrates such as tadpoles, and will even eat small fishes such as minnows and sticklebacks. The nymph of the locust (*Locusta migratoria*) is called a 'hopper' because of the way that it moves. The hoppers slowly migrate across parts of Africa, Asia and India, eating much of the vegetation and crops in their path, and hence they are feared and hated by the farmers of these regions.

Nymphs moult several times as they grow, and during the last few sheddings of the skin wings develop. From the final moult an adult insect emerges.

The most complex type of life history is seen in the more highly evolved orders such as those of the flies, beetles, bees and butterflies. It involves four stages.

The numerous eggs hatch out into tiny larvae, which eat almo continuously, initially to mak up for the small amount of yo. that their tiny eggs held. The f maggot and the caterpillars moths and butterflies are larv. stages. These young can caus terrible damage for humans. Th cabbage white butterfly larva ca cause havoc with cabbage crop clothes moth larvae eat holes c damage fibres in garments ar furnishings, whilst some beet larvae are feared by museun because of the damage that the cause, not only to the buildin timbers but also to the preciou contents and specimens.

Many larvae eat an enormou amount of food, not simply t store up energy, but because th food is often not very rich i nutrients. Some larval stages ma

Below: *Two male stag beetles* (Lucanus elephas) *have locked their antler-like mandibles in a fight. The jaw muscles ar too weak to move the massive 'antlers' very much so it is most unlikely they could hurt any animal with them.*

Right: A female stag beetle takes flight, showing the first pair of wings modified as wing cases, whilst the thin second pair do the actual flying.

live actually in their food: for example, some beetle and fly larvae live in dung. The stag beetle larvae live in rotten timber, such as oak logs, for anything up to 5 years, and so obviously there is not much nourishment available if it takes so long for them to mature, despite the great amount that they eat.

Eventually all insect larvae find a site where they will be protected from the worst of the weather and predators. They then turn into the third stage of the life history – the pupa. This is the fixed stage and is seen in the very familiar butterfly chrysalis. Within the thin shell of the pupal case the animal changes from a larva into an adult. This dramatic conversion from one stage into another is known as metamorphosis.

The adult in all life histories is responsible for reproduction and in some cases may lay thousands of eggs. Often the active adult life is very short – the stag beetle (*Lucanus cervus*) mentioned earlier lives for only 2 months whilst the adult mayfly lives for only a few hours.

Whatever type of life history an insect displays, there is often a great difference between the food of the young and that of the parent. Many butterflies have larvae that feed on leaves whilst the adult sips nectar. This is a great advantage as it prevents competition within a single species. Contrast this with the situation in mammals where the new-born feed firstly from the mother's milk and then on the same food as the parents.

Types of insect

Due to the large numbers involved and the many different orders into which this mass of insects have been classified, it is only possible to describe some of the many important groups.

Beetles

(ORDER COLEOPTERA)

The beetles form the largest group of insects with nearly 300,000 different species. This may be compared with the vertebrate animals such as fishes, amphibians, reptiles, birds and mammals which together only amount to 43,000 species.

The most noticeable feature of the beetles is the pair of wing cases, or *elytra,* which are developed from the first pair of wings. Not only do the elytra provide protection for the delicate second pair of flying wings, but they usually protect most of the abdomen as well. The beetles' success is partly attributed to their protective elytra, and partly to their ability to feed on an amazingly varied diet. The different species are known to eat almost every conceivable food material, from living vegetation, humus, organic soils, carrion, dung, timber, other animals and even specialized materials such as silk. With such a variety of diet it is not surprising that beetles are found worldwide in most types of habitat.

The ladybirds are very well known. They are found in every inhabited continent and they form a very important group of predators. They are best known and respected for their ravages of the aphids or greenflies. Ladybirds eat aphids in vast numbers and have even been raised on special ladybird farms in order to sell them to horticulturists who feel they could do with additional help in fighting the pests of their crops. Many millions of ladybirds have been released into the orange groves of California, for example.

The weevils are almost as much hated and feared as the ladybirds are loved. They are a very widespread and commercially important group of herbivores. They are often found attacking fruits and seeds. The gorse weevil, as the name implies, attacks gorse seeds and has been introduced deliberately into New Zealand to help check the spread of the wild gorse. However this useful aspect of weevil behaviour is very much overshadowed by the tremendous damage that the group does to man's crops. The cotton boll weevil has become notorious, and is commemorated in the folk songs of the United States. The damage that this insect caused to the cotton crops was so great that the dependence of the economy of the southern states of America on the two great crops of cotton and tobacco had to be ended, and agricultural production diversified into other branches. Root crops, such as sugar beet, carrot and turnip, all suffer from attack by weevils, and the grain weevil presents one of the greatest problems facing any farmer wishing to store cereals.

The scarab beetles are famed

for their splendid colours, their large size, and their valuable help in burying dung. The balls of dung are buried as a food store for their young, but this action obviously aids the soil fertility.

The largest beetles in the world, the African Goliath beetles, may reach 10 centimetres in length. They are closely followed in size by the grotesque rhinoceros

Above: *Unlike the ladybirds, it is hard to find anything pleasant in the weevils' favour. They are the largest group of beetles and probably do the most harm, feeding on man's crops and stored grain.*

beetles. The British stagbeetles may grow as large as 5 centimetres. Fortunately their fearsome jaws are harmless to man for the jaw muscles are far too weak to cause a painful bite.

Left: *A scarab beetle rolling a ball of dung which it will bury as a food store for its young. Most scarab beetles have a metallic lustre to their exoskeletons; some are used to make jewellery.*

Moths and butterflies

(ORDER LEPIDOPTERA)

It is difficult to state the exact difference between moths and butterflies. However, most moths fly at night, have feathery antennae and fold their wings flat across their back, whilst most butterflies fly during the day, have antennae with swollen ends and fold their wings vertically above their backs. These points are very general and exceptions can be found to all three features.

True flies

(ORDER DIPTERA)

This very large and important group can be distinguished from all other insects as they have only a single pair of wings, the second pair being reduced to stick-like balancing organs.

Mosquitoes are notorious for their irritating and sometimes painful bites. The male mosquitoes are specialized for feeding on plant sap and nectar, so that it is only the females that pierce animals to suck up the blood. Although their bites are a nuis-

ance, the wound itself is not dangerous, but danger does lie in the possibility that the victim of the mosquito will be infected with some very unpleasant disease when the mosquito injects her saliva. Yellow fever and malaria are two very well known and extremely dangerous diseases that are transmitted by mosquito bites.

As all mosquitoes breed in water, it is by spraying ponds, lakes and puddles, and by improving drainage that man can best eliminate these pests. However, despite considerable success since the second world war, it has

become clear in the last few years that man's optimism concerning the eradication of malaria has been unfounded, as the mosquitoes are again spreading and building up their numbers.

Houseflies and their close relatives, the blue- and greenbottles, feed on a wide range of man's food and decaying materials. The larvae, the maggots, live in their food and unlike caterpillars have no limbs. Due to their indiscriminate feeding on dung, carrion or human food, the houseflies spread disease by transmitting bacteria either on their mouthparts or their legs.

The best way of getting rid of these pests is to limit their food supply by keeping all rubbish under control and food under cover, and by ensuring that sanitation is generally of a high standard. Flies have been destroyed in vast numbers using a variety of sprays and insecticides, but there are many strains that are now resistant to sprays, and

Below: *Bluebottles are a cause of irritation to people in the summer when they drone around the windows. In fact male bluebottles prefer to feed on flowers, only females usually entering houses.*

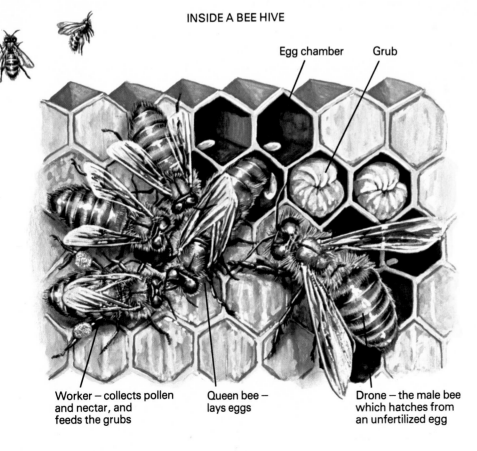

INSIDE A BEE HIVE

Egg chamber   Grub

Worker — collects pollen
and nectar, and
feeds the grubs

Queen bee —
lays eggs

Drone — the male bee
which hatches from
an unfertilized egg

hence the measures concerning sanitation and general hygiene will be by far the most important in the long run.

Troublesome relatives of the houseflies, the tabanids, can inflict very nasty wounds and cause considerable discomfort to man and his stock.

The tsetse fly (*Glossina* species) may transmit the disease of sleeping sickness when it bites man and his stock, carrying the disease from the native fauna. Vast areas of the African continent have been made untenable to man's stock because of this fly and the disease which it carries.

Wasps, bees and ants

(ORDER HYMENOPTERA)

This is a large group of insects, many of which are colonial, acting together for their mutual benefit. Thus the honeybee colony centres around the large

ft: Matabele ants can be seen returning
m a raid carrying a variety of dead and
ing prey. These foraging ants show
at organization in their savage search.

een that is responsible for egg
ying, whilst the drones' only job
to fertilize the queen's eggs. All
e other jobs are carried out by
e hard working sterile females,
e workers.

Ants similarly have complex
olonies, often with many spec-
lized types of individual. Some
arts of the world have ant col-
ies that are nomadic, that is
ey keep on moving from place
place. Examples are the army
ts of South America and the
iver ants of Africa. These parti-
ular species are savage carni-
ores and will attack almost any
ving animal unfortunate en-
igh to be in their path.

Whereas most ants build their
ests underground from soil,
me bees use a special form of
ax in the making of their hives,
d some wasps build delicate
ests from chewed wood pulp.

The resulting dried material
looks and feels like coarse paper.

Fleas and lice

(ORDERS SIPHONAPTERA
SIPHUNCULATA, PSOCOPTERA,
AND MALLOPHAGA)
Fleas and lice are parasites that
are found on the outsides of a
wide range of hosts. They are not
in the same group but are often
thought of together. Both types
are wingless insects and infect
mainly mammals and birds.

Fleas tend to be found on birds,
such as woodpeckers and sand-
martins, that nest in holes, and on
mammals that live in lairs or
dens. Man, in fact, is one of the
very few primates to suffer flea
infestation.

Again, the problem with these
parasites is not just the irritation
that they cause and the blood that
they take, but the possibility that
they may infect the host with
some disease. Bubonic plague, or
the 'black death', of mediaeval
times was transmitted by rat fleas

from carrier rats to humans.

Typhus is a terrible disease
that appears in dirty crowded
communities, and it can be spread
by lice.

Grasshoppers, locusts and cockroaches

(ORDER ORTHOPTERA)
This is a group of voracious herbivores and scavengers. The grasshoppers and crickets are harmless occupants of gardens and
meadows, but the locusts are
devastating pests, causing untold
misery and ruin with their eating
of man's crops. Cockroaches are
found in and around buildings
and they eat stored food and
rubbish.

Stick insects, familiar to many
children as school laboratory
'pets', are close relatives of the
grasshoppers and exhibit some of
the finest camouflage in the insect class.

Below: The fringed legs and body of the
giant stick insect camouflage it against
stems.

Echinoderms (PHYLUM ECHINODERMATA)

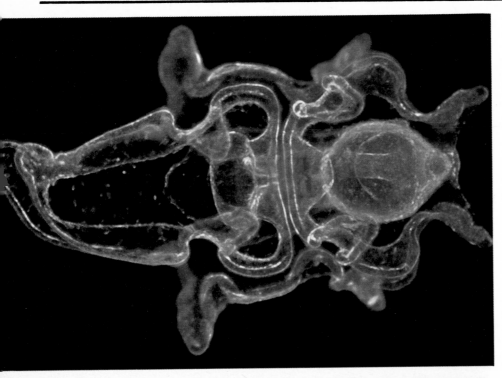

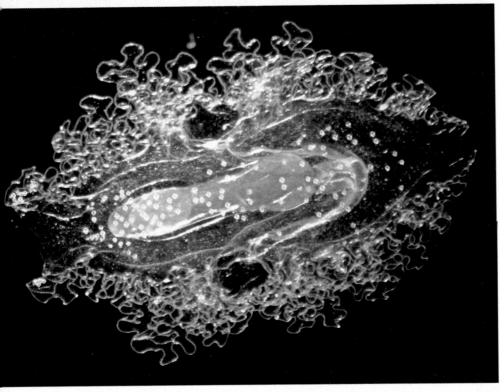

The echinoderms are an important group of invertebrate animals because their larvae are similar to those of the hemichordates (see page 93). Hemichordates are in turn linked to animals with backbones.

The larval forms vary depending upon which class of echinoderms they belong to, but all of them are active, moving busily in the upper layers of the oceans. They move using small hair-like structures, and form an important part of the plankton referred to in the chapter on the seas.

Although the larvae are active and can be seen to have a left and right side, the adults are very slow moving, sometimes stationary, and have a radially symmetrical body plan. The body is divided into 5 parts around a central axis. The echinoderms are all marine, living on the sea bed under rocks or crawling over coral reefs.

It is often difficult to make out an echinoderm when looking into a deep rock pool as they tend to crawl into cracks or half under rocks, and their surface is often covered in spines or outgrowths that break up their outline and tend to blend with the rocks and seaweeds. In addition to the spines, echinoderms have tiny tubes of skin called tube feet sticking out through their armour plating. The tube feet are used to move the animals, to act as suckers, to hold food or rock and even to breathe through.

Over 5,000 species of echinoderms have been recorded and they are divided into 5 classes: the starfish (Asteroidea); the brittlestars (Ophiuroidea); the sea urchins (Echinoidea); sea cucumbers (Holothuroidea) and the sea lilies and feather stars which belong to the Crinoidea.

Top: *This curiously-shaped larva is an immature starfish. The shining lines are rows of fine hair-like cilia that propel the animal through the seas. Cilia also trap food.*

Above: *This larval sea cucumber has even more twisted rows of cilia. The central body mass has already formed the elongated shape of the adult. Both these larvae show the plankton's familiar transparency.*

Starfishes

These are found in all the oceans of the world but, like all echinoderms, they occur most commonly in the warm waters of the Pacific and Indian Oceans. Starfish are easily recognised by their body shape – most have 5 arms developed from a central body mass which has no head. The mouth is on the lower surface, and there is a groove running along the undersurface of each arm which joins the other grooves in the mouth region. These grooves are full of the very active tube feet.

Starfish are often a yellowish brown colour although the tropical species tend to be brighter with splashes of vivid scarlet and yolk yellow.

Right: *Starfish have many fine tube feet which extend from their body surface. These tube feet help in locomotion, feeding and respiration. The largest tubes are found along the grooves under the 5 arms.*

Carnivorous starfishes

All starfish are carnivores, although many will also eat dead material if they find it. The long muscular arms of the starfish are useful in holding prey such as molluscs, small crustacea and the bottom-living bristleworms. Some of them are capable of pushing their stomach out through their mouth and this wraps around the prey and starts to digest it. When the prey is partly broken down, the stomach is sucked back into its normal place inside the animal, together with the softened bits of the latest meal. One of the best known starfish today the crown-of-thorns starfish (*Acanthaster planci*) that feeds on living coral. This large and voracious animal has caused considerable damage to the Great Barrier Reef off north east Australia.

Right: *The crown-of-thorns starfish has caused great problems in recent years by eating the living polyps of coral reefs. The destruction of the reefs causes changes in the food chains and water currents.*

Brittlestars

(CLASS OPHIUROIDEA)

Brittlestars are very similar to starfish but the central body is far smaller and the limbs are long and thin. They get their common name from their odd habit of shedding limbs when attacked, the detached arm continuing to wriggle about, presumably acting as a decoy, whilst its recent owner makes a very slow escape! Brittlestars, like starfish, are cap-able of regrowing lost limbs and hence can easily repair their losses. Brittlestars are less often seen than starfish as many are nocturnal and they spend much of their time buried in mud, under rocks or in crevices. Because they are so slow they cannot hunt very actively, and live on particles of plant and animal material drift-ing in the sea and lying on the sea bed.

Because of their shape they are less robust than starfish and it is not surprising that they are not as common around the coasts as

Above: *These 3 brittlestars are typical of their class. They show the compact body mass and the very elongated narrow arms. The tips of the arms are very active and can be seen writhing around. It is the tips that pull the animals into crevices but they stay there because of the spines.*

their tougher relatives. However they do extend far out beyond the continental shelf and slopes some have even been fished up from depths of over 6,000 metres. Deep down in the ocean there is not so much activity, currents are slower and there are no waves to tumble delicate animals about.

Sea urchins

(CLASS ECHINOIDEA)

The urchins are the echinoderms found near the shore and hence the battered remains of their spherical shells can often be found at the edge of the tide. Although they appear bulky and awkward with their forest of spines, many are capable of burying themselves in mud with remarkable speed. The burrowing habit is very common in sea urchins and some are known to drill holes in packed sand, soft limestone and even hard rock.

The spines are a very good defence against enemies whilst being of help in movement. The spines work like stilts and are assisted by dozens of long tube feet which are used as suckers to steady and pull the animal. This unlikely combination of spines and suckers has enabled many sea urchins to climb rocks, jetties and even the glass sides of aquaria.

The mouth is again on the lower surface and is equipped with 5 large teeth, which, together with the extensions for muscle attachment resemble an old-fashioned oil lamp, hence the name of Aristotle's lantern that is given to dried specimens of mouthparts. Carrion and sea weed both play an important part in the diet of these scavengers.

The dried shells of these animals are often sold as colourful souvenirs in shops around the seaside resorts of the world.

Right: *The hat-pin sea urchin* (Diadema setosum) *is a tropical species which shows the long tapering spines, typical of the sea urchins. In some Pacific species the spines may reach 40 centimetres in length! The sea urchin walks on the spines as if on stilts, but the spines are moved by muscles which permit quite delicate manipulation. The tube feet can just be seen between some of the spines. Some spines contain chemicals that may cause intense irritation to human flesh.*

SEA URCHIN STRUCTURE

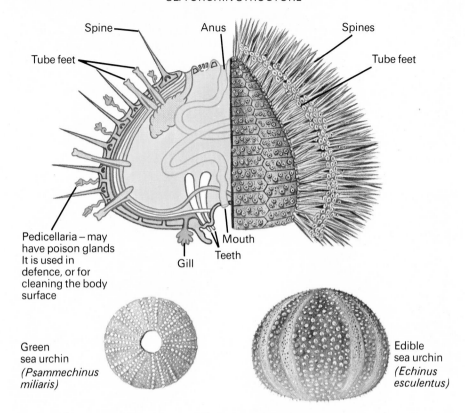

Spine

Anus

Spines

Tube feet

Tube feet

Pedicellaria – may have poison glands It is used in defence, or for cleaning the body surface

Mouth

Teeth

Gill

Green sea urchin *(Psammechinus miliaris)*

Edible sea urchin *(Echinus esculentus)*

SEA URCHIN SHELLS

Sea-cucumbers

(CLASS HOLOTHUROIDEA)

These tubular echinoderms lack spines and rigid shells, the skin being leathery and protected by small chalky plates called spicules. Sea-cucumbers are found mainly in warm waters although some may be at great depths. They are very sluggish animals feeding on the fine particles of animal and plant remains in the mud and sand. They feed using a ring of tentacles around the mouth.

They have a habit of ejecting their gut when threatened, and then abandoning the mass of twisting tubes whilst creeping away to grow a new set.

Sea-lilies and feather-stars

(CLASS CRINOIDEA)

Sea-lilies are found in deep waters where there is little water movement. They are fixed by a 'stalk' to the bottom, with a cup-like body on top of the stalk. Around the rim of the cup are 5 arms which are often divided into smaller 'fingers'. They feed on very fine particles that they sieve from the water.

Feather-stars have branched delicate limbs like the sea lilies and are stalked in their young stages but become capable of free movement in later life, although they are never active animals.

Above, left: *A tropical white sea-cucumber showing the delicate branching tentacles around the mouth. The tube feet are arranged in rows. The feeding tentacles are modified tube feet. It is difficult to see the connection between the delicate, transparent larva on page 86 and this heavy animal.*
Left: *Feather-stars of the tropics feed by waving their branching tentacles to trap food.*

Bryozoans (PHYLUM ECTOPROCTA)

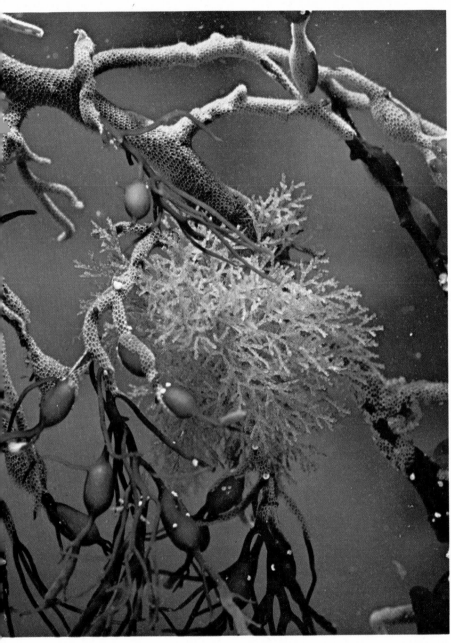

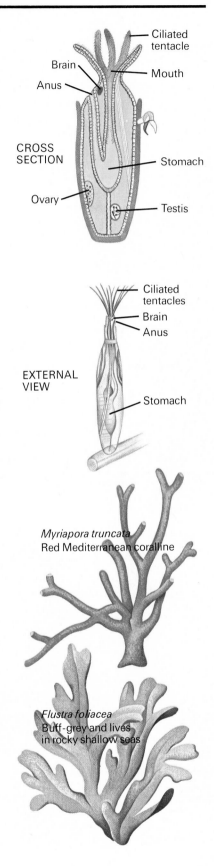

CROSS SECTION

- Ciliated tentacle
- Brain
- Anus
- Mouth
- Stomach
- Ovary
- Testis

EXTERNAL VIEW

- Ciliated tentacles
- Brain
- Anus
- Stomach

Myriapora truncata
Red Mediterranean coralline

Flustra foliacea
Buff-grey and lives in rocky shallow seas

These form a large group of small colonial animals found mainly in the sea but with a few freshwater forms. Each individual looks like a miniature sea-anemone or a moss plant and so the bryozoans are often called 'moss animals', but closer investigation shows that despite the tentacles and sedentary habit, each animal in fact has an internal structure closer to a worm than an ane-

Above: *The encrusting bryozoans on this seaweed are often called the 'sea mats'. There are 2 main growth forms, both shown here, the branching type is* Scrupocellaria scruposa *and the white encrusting one is* Electra pilosa.

mone. When thousands of these cluster together they form a dense 'mat'; with their body walls thickened with lime secreted by themselves they resemble a layer of coral.

Arrow worms (PHYLUM CHAETOGNATHA)

ARROW WORM STRUCTURE

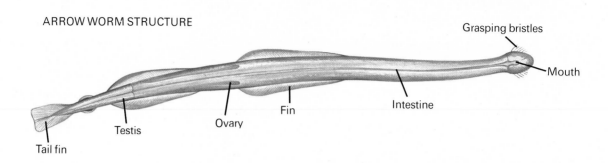

Grasping bristles

Mouth

Intestine

Fin

Ovary

Testis

Tail fin

The arrow worms form another marine group found amongst the drifting plankton of the oceans. However, unlike the slow-moving ciliated larvae of echinoderms and worms, the arrow worms are voracious hunters swallowing their prey whole. They get their common name from their long thin shape and swift darting movements. Their movements are very difficult to follow as they exhibit a common planktonic characteristic – they are transparent and hence show the finest form of camouflage.

The hunger of these small animals, between 2 to 10 centimetres long, is frightening as they will swallow crustacean larvae and small fish their own size. Reports of young herring 5 centimetres long being swallowed entire, and digested at leisure, have been very well authenticated.

Their significance as a group lies in two features; firstly they are important predators in the food web of the plankton, and secondly their internal structure is rather confusing having some features similar to those of the echinoderms and simple chordates, but the relationship between the arrow worms and these other groups is uncertain.

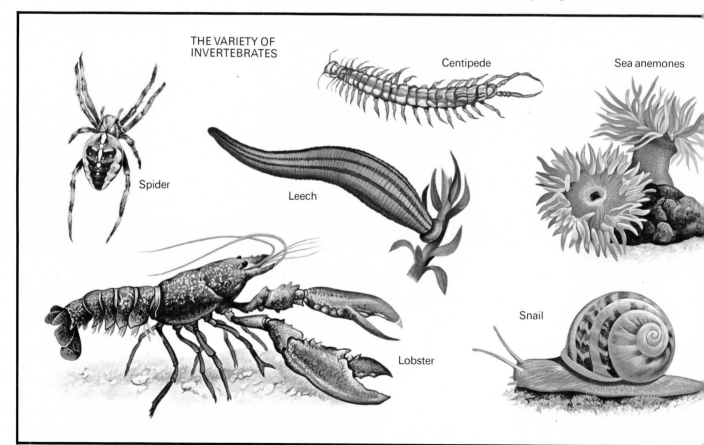

THE VARIETY OF INVERTEBRATES

Centipede

Sea anemones

Spider

Leech

Lobster

Snail

Acorn worms

(PHYLUM HEMICHORDATA)

This small group of marine animals is amongst the most primitive of all chordate-like animals, that is, animals having some form of backbone. Looking at a typical hemichordate such as *Balanoglossus,* the acorn worm, it is easy to think it is a worm. Its body is tubular and soft and it even lives and feeds in a similar way to many of the bristle worms. *Balanoglossus* is found in burrows in the mud and sand of shallow coastal waters, and feeds by sticking its head region out of its hole and pulling in mud and bits of silt from which the edible organic remains are removed.

Acorn worms get their name from a domed structure – the proboscis – that extends forwards and can be pulled back within the collar that extends around the mouth.

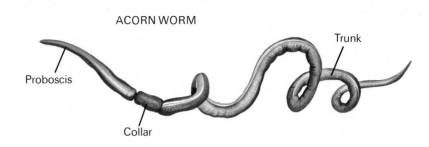

ACORN WORM

Proboscis

Collar

Trunk

The real importance of this group to the biologist lies in their tiny larvae which drift with the plankton in different parts of the world. The larva, called a *tornaria,* has a very simple gut and possesses a single band of fine hair-like cilia running around its body. The form of the tornaria very closely resembles echinoderm larvae, whilst still showing some similarities to the larval stages of the worms and molluscs. The adult worm has a notochord (a primitive backbone) comparable with that of the chordates, and hence it is thought that this group provides an evolutionary link between the echinoderms and the chordates.

Below: *This acorn worm lives in very shallow waters amongst silt and organic rubbish upon which it feeds.*

Butterfly

Vorticella

Chordates (Phylum Chordata)

The chordates contain the well-known animals with backbones, the vertebrates. They include the fishes, the amphibians, the reptiles and the birds. There are 2 groups, however, that do not have backbones but possess certain chordate distinguishing characteristics. These are the tunicates and the lancelets. Both these groups at some time in their life history have a notochord, a dorsal hollow nerve cord, and pharyngeal or gill clefts, all chordate features.

Tunicates

(Sub-phylum Urochordata)

Adult tunicates do not look at all like other chordates. Most of the 1,200 species are barrel-shaped

Above: The tubular form of the sea squirts is shown clearly, the main opening at the top with its crinkly edges can be seen beneath the shrimp. Water entering is sieved and finally leaves via the side openings. Encrusting animals do no harm to them.

creatures attached at one end to a rock or similar object. Only the larval stage shows the distinct chordate characteristics. The larva looks like a miniature tad

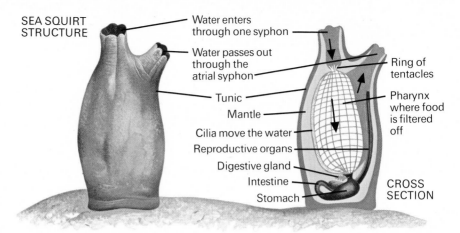

SEA SQUIRT STRUCTURE

Water enters through one syphon

Water passes out through the atrial syphon

Tunic

Mantle

Cilia move the water

Reproductive organs

Digestive gland

Intestine

Stomach

Ring of tentacles

Pharynx where food is filtered off

CROSS SECTION

ole. One group, the sea squirts or ascidians, are fixed tunicates that are common marine creatures throughout the world. They are motionless, and at first examination look rather like a shapeless lump of matter covered in a rough leathery skin, hence the name tunicate. At the top of the sea squirt is an opening into which water is drawn. Inside the tunic is a large barrel-shaped strainer through which the water passes and it flows out of a second opening at the side of the animal. Food particles are separated by this strainer and are passed into the stomach and then the intestine. There is a very simple nervous system and no trace of a brain, spinal cord or notochord. However, the straining barrel not only filters food but also allows the animal to breath. It is a complicated gill mechanism and typically chordate in structure.

The sea squirt larva shows the other chordate characteristics. Looking like a tadpole in shape, with a large head and long slim tail, this small animal swims about. In the tail is a well-developed notochord and a typical chordate nerve cord as well. When it finds a suitable resting place such as a rock, shell, piling or ship bottom, it becomes attached in the head region. It abandons its active life, its gill barrel expands and the tail with nerve cord and notochord disappears.

The other tunicates are the free-swimming salps that look similar to sea squirts, but the 30 or so species live in the plankton of mainly tropical and subtropical seas. The other group, the larvaceans, are tiny, transparent, free-floating animals that filter minute planktonic animals from the water. Some adults remain as larvae but can reproduce.

Zoologists believe that tunicates departed early from the chordate line of evolution. They show no signs of segmentation and at no stage is there a coelom (body cavity), which has been lost in their evolution. In higher chordates segmentation probably evolved much later.

Lancelets

(SUB-PHYLUM CEPHALOCHORDATA)

The 20 or so species of lancelets are small segmented animals that live close to the shore on sandy seabeds, mainly in tropical seas. They are fish-like in appearance but in structure are much more primitive than any fish. There are no paired fins or limbs, no jaws, bones or cartilages or even a backbone. They do have a notochord and dorsal nerve cord. They spend much of their lives partly buried, tail first, in the sand drawing in water through their mouth and filtering tiny particles by their gill clefts. They reproduce sexually and the sexes are separate.

It is possible that the lancelets are fairly close to the primitive types from which the backboned vertebrates have evolved, although lancelets living today have some specialized features.

Below: *This marine lancelet has been photographed against a dark background to show the fine tentacle-like oral cirri at the head end (right). These fine structures sieve out sand particles as they draw in water which passes out over filtering gills.*

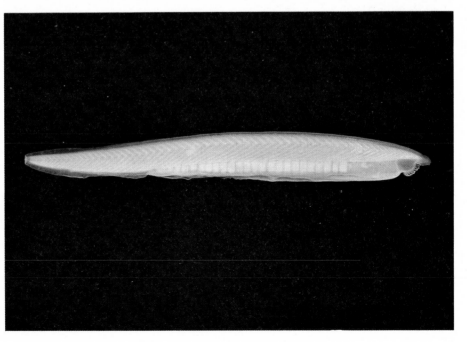

Fishes

Fishes are the oldest vertebrate group, and the most numerous. Totally aquatic fishes are to be found almost everywhere that there is permanent water – from fast-flowing, bubbling mountain streams, to deep-sea waters where there are tremendous pressures of water and no light. Although fishes are cold-blooded, there are species adapted to living in quite hot waters or even freezing waters. Some fishes can withstand living in waters that occasionally dry up, having evolved internal organs that allow them to breathe the oxygen of the air. Each species is usually adapted to living in a certain habitat and will rarely venture into a different one. However, their range is amazing. There are fishes that, apart from simply swimming, can walk, crawl and glide, and others that can look like seaweed, sand or rocks; wear armour-plating; inject venom; or change their colour to match their surroundings. Thus they have evolved to fit almost every niche which is open to them.

The evolutionary history of fishes stretches back over 420 million years, and the group has a good fossil record, so that many pieces of their past have been fitted together.

The earliest fishes did not have bony jaws, but most members had heavy bony armour that covered and protected their bodies. This probably meant that they were poor swimmers. One extremely heavy, armoured fish was *Cephalaspis*, and it is thought that it fed on food sucked up from the muddy beds of streams. These jawless fishes (CLASS AGNATHA) are represented today by the soft-bodied lampreys and hagfishes.

The earliest fishes with jaws were the sharks, and they appeared about 370 million years ago. Sharks and rays have skeletons of cartilage, a gristle-like substance that does not fossilize well. However, although full skeletons are very rare, teeth and fin spines are often found as fossils, and a picture of the early sharks can be built up from these. One of the earliest members was *Cladoselache*, which lived in the late Devonian period, some 350 million years ago. It was a marine fish and grew up to 1.2 metres in length. As sharks evolved, they developed jaws with teeth adapted for killing and slicing prey, or with low, flat teeth suitable for crushing shellfish. Sharks are the major predators in the fishes' world.

The most successful fishes appeared in the Devonian period, these being the advanced bony fishes (CLASS OSTEICHTHYES). They radiated to become one of the most diverse vertebrate groups, and today over 20,000 species are found. This number is greater than the total of all other vertebrate species combined. The ray-finned fishes or teleosts are the most successful and commonest bony fishes, and they include salmon, sticklebacks, mackerel, herring, trout and cod, to name but a few. Today the teleosts are at their peak in the seas and fresh waters of the world. They have evolved to fit into every available niche.

The lobe-finned fishes (ORDER CROSSOPTERYGII) are very interesting in that they probably include the ancestors of the land vertebrates. About 360 million years ago the lobe-finned *Eusthenopteron* lived. The base of the paired fins was fleshy and muscular. Strong fin rays radiated from this lobe and could probably be moved. It is thought that this development of the fin allowed the fish to 'walk' on land, and the skeleton of this fish and its relatives is very like that of the early amphibians.

Right: *Fishes were the first group of vertebrates to evolve. Today they dominate the waters of the world. The majority of fish have muscular, streamlined bodies for high speed swimming as shown by this shoal of Caranyx.*

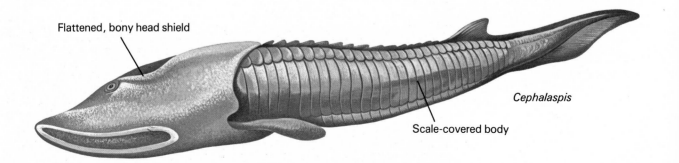

Flattened, bony head shield

Scale-covered body

Cephalaspis

Jawless fishes – scavengers and parasites

(CLASS AGNATHA)

The only surviving members of the earliest vertebrate group are some 45 species of hagfish and lamprey. As well as being jawless they have no scales and they are either parasitic on other fishes or are scavengers.

Lampreys

(ORDER PETROMYZONTIA)

Lampreys have a circular, sucking mouth with horny teeth and a rasping tongue. With this equipment they cling on to other fishes and rasp at the flesh, draining the host of its blood. They live in the temperate seas and fresh waters of the northern hemisphere.

The sea lamprey (*Petromyzon marinus*), which is found on both sides of the Atlantic, is probably

the best known. Although it spends its adult life at sea, it enters freshwater rivers to spawn. It begins to migrate in winter and by spring the lamprey has struggled up a river. It can swim strongly, although it lacks paired fins, and can get over almost vertical rocks by using its sucker

Above: *Lampreys and hagfishes, such as this river lamprey, are parasitic scavengers with no scales or jaws. The river lamprey attaches itself by suckers to the side of a host fish and then rasps through skin and flesh to drain the animal of its blood.*

mouth. The male begins to build a nest by moving pebbles, in his mouth, to make a barricade. On

LAMPREY LIFE HISTORY

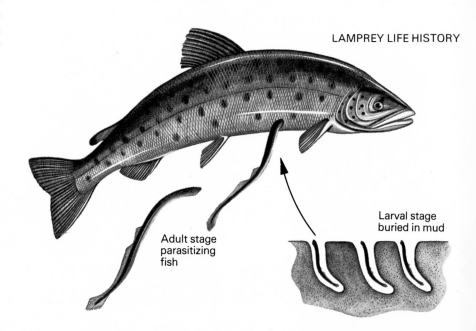

Adult stage parasitizing fish

Larval stage buried in mud

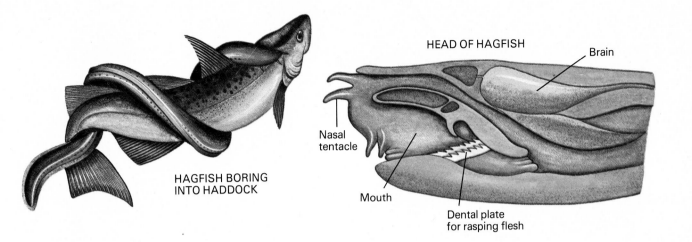

HEAD OF HAGFISH

Brain

Nasal tentacle

Mouth

Dental plate for rasping flesh

HAGFISH BORING INTO HADDOCK

the upstream side of this, a depression is excavated, and this is where the female will lay her eggs. The female helps in the later stages. A pair have been seen working together in order to remove large pebbles. After spawning the adults drift downstream and die.

The lamprey eggs are 1 millimetre in diameter and hatch in about 14 days. It was once thought that the larvae were a distinct species. Because they look so unlike their parents they are called *Ammocoetes branchialis*. They are worm-like, with rudimentary eyes buried beneath the naked skin. The horse-shoe-shaped mouth has a small lower lip, and the 'shoe' is formed by the upper lip. There are no teeth, but at the entrance to the mouth there are a number of fringed barbels, forming a perfect strainer. After staying in the nest for about 1 month the larvae wander downstream until they find a sandy or muddy bed. Here they bury themselves and remain in their tube for 3 to 4 years. Quite blind, they feed by filtering minute organic matter from their surroundings. Between 3 and 5 years of age, over a period of about 8 weeks, they metamorphose into adults. Eyes appear, the mouth becomes sucker-like, and teeth and rasp develop. Internally many changes are also taking place, until silvery adult lampreys are identifiable. These go down to the sea and lead their adult lives parasitizing fishes.

Hagfishes

(ORDER MYXINOIDEA)

A hagfish has been described as a fish with 4 hearts, 1 nostril, no jaws and no stomach. This rather weird description is quite correct for this degenerate descendant of the jawless fishes. The 15 species all live at depths of 60 metres or so in temperate seas of both hemispheres. Their slimy, eel-like bodies are up to 60 centimetres long with vestigial eyes and a row of gill openings visible. Running along the body are small openings to slime glands. These glands secrete a glutinous mass of slime cobwebs, which makes the fish very slippery. This is the reason why they are sometimes called slime eels or glutinous hags.

During the day hagfishes remain buried in sand, gravel, or mud beds. At night they emerge to scavenge on dead animals or organic waste, also taking worms and crustaceans. They are better known, however, for rasping into the bodies of dying fishes, eating away until all that is left is little more than skin and bones.

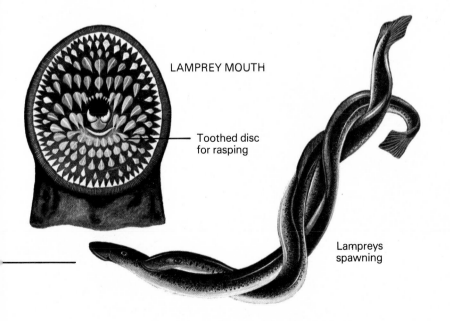

LAMPREY MOUTH

Toothed disc for rasping

Lampreys spawning

Cartilaginous fishes

(CLASS CHONDRICHTHYES)

The cartilaginous fishes are jaw-ed animals with skeletons composed primarily of cartilage, with mosaics of small bony plates as strengthening parts. They are grouped in the class Chondrichthyes and include sharks, rays, skates and chimaeras. The mouths are adapted for biting and slicing, or crushing, with a band of teeth fitted in each jaw. The paired fins act as stabilizers and steering organs in sharks, while in rays and skates they are the main swimming aid. The cartilaginous fishes do not have the swim bladder which helps most fishes to keep buoyant. Instead the flattened head, pectoral fins and tail help to give lift. The gills, 4 to 7 in number, open to the outside individually. These gill slits can be seen behind the head region.

The sense of smell is very well developed and the lateral line system is extremely sensitive. This system consists of a network of pressure-sensitive organs found on the head and along the sides of the body. It detects the vibrations in the sea, and the nervous system links them up to the brain. Just behind the eye is a hole, called the spiracle. Spiracles have evolved from a former pair of gill clefts but no longer serve a respiratory function. However, in many species water is taken in via these openings to be passed over the gills and out of the gill slits.

It is easy to distinguish a male from a female in this group. The male has a pair of stiff, finger-like 'claspers' attached to the pelvic fins. These are used to help the male carry out internal fertilization. Many sharks and rays bear their young alive, that is, the young hatch from the eggs inside the female's body, but others such as the Port Jackson shark, the dogfishes and some of the rays, lay eggs. Each egg is encased in a strong horny capsule, sometimes called a mermaid's purse, sailor's purse or mermaid's pin-box. They may frequently be picked up on the seashore after storms.

Below: *Curling tendrils attach the egg case of a nurse hound fish to a stem. Inside the egg case can be seen the developing shark with its long tail, bulging eyes and the dark yolk sac that provides it with nourishment.*

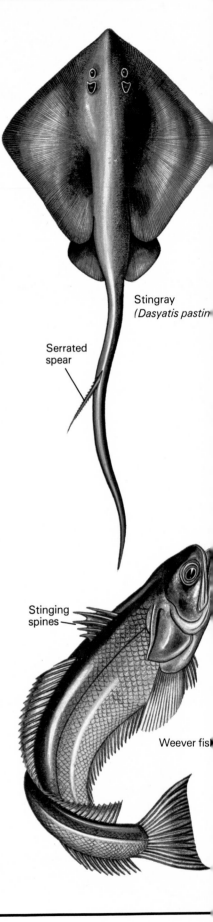

Stingray
(*Dasyatis pastin*

Serrated spear

Stinging spines

Weever fis

Left: *Ugly and most venomous, the stonefish lies concealed in shallow tropical waters of the Indo-Pacific ocean. The sharp spines along the back are linked to poison glands.*

Spears, stings and shocks

The majority of fishes are quiet, inoffensive and either camouflaged or brightly coloured for various reasons. However, there are some fishes that are well equipped with powerful weapons of defence.

Stingrays have their weapon at the base of the tail in the form of a long serrated spear. Fishermen fear treading on one because, when stimulated, the ray curls the tail over its back and jabs the sting into the victim's leg. In Australian seas, some stingrays are almost 2 metres across the body, and wounds from their stings have caused several deaths.

The stonefishes (*Synacceja* species), of Australian tropical waters and most of the Indian and Pacific Oceans, are shallow-dwelling fishes, beautifully concealed with their warty, brown blotched bodies. On the dorsal fin is a series of short, very sharp spines each linked to venom glands. When a victim is stabbed by these spines, the wound is immediately flooded with venom. Apart from the extreme pain of the wound, infections are often started.

The most venomous fish of British and European seas is the weever (*Trachinus vipera*), which, although small at 13 centimetres, is distinctly dangerous. The dorsal fin spines and the spines on the head can produce very painful burning stings. Sailors have been in such agony from these stings that they have attempted to throw themselves overboard.

Unique among vertebrates is the remarkable electrical property of certain fish species. There are about 250 different kinds of fishes that possess electric organs. Among the best known are the torpedoes (ray-like fishes of tropical and temperate waters), electric eels of South America (which are not true eels), the African electric catfish and the marine stargazers. Weak electric power was probably evolved as a means of signalling to one another as in the rays, or as a navigational aid, as in the elephant-snout fishes of African fresh waters. Some fishes have developed their electrical properties to such an extent that the shock produced is powerful enough to assist in catching food or warding off enemies. The electric eel of South America can produce over 500 volts at a time, sufficient to knock down into the water a wading man who touches the fish. The electric catfish can produce a shock of between 350 and 450 volts, while a torpedo can only manage about 40 volts.

The electric eel and torpedo have been seen to stun small fishes before devouring them. The electric catfish probably uses its electrical powers for defence.

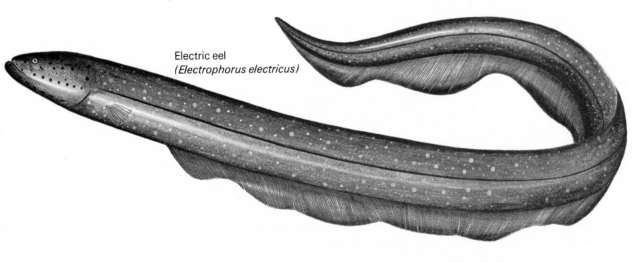

Electric eel
(*Electrophorus electricus*)

Hammerhead shark

Sharks

(ORDER SELACHII)

Most sharks are rather long-bodied, streamlined creatures with eyes and gill-openings on the side of the head. Most of the 200 species are predators, feeding on smaller bony fishes. Only 12 or so species are dangerous to man. As well as the paired fins, they have 1 or 2 dorsal fins on the back and 1 anal fin just in front of the tail fin. The body is covered with sharp, pointed scales called *denticles* which point backwards. The skin feels fairly smooth if rubbed towards the tail, but very rough when rubbed in the opposite direction.

Sharks are found in most seas of the world and a few species enter fresh water. The most primitive are the rare, very slender frilled sharks (*Chlamydoselache* species) of North American and Japanese waters, and the more common cow sharks (*Hexanchus* species). These sharks have 6 or 7 gill slits, while the more advanced sharks have only 5 gill slits. These include the sand sharks (*Carcharias* species) and the mackerel shark family which contains some of the world's most dangerous fishes. They are typically shark-like in appearance with numerous sharp, wicked-looking teeth protruding from the mouth. Included in this family is

the great white shark (*Carcharodon carcharias*), commonly called the man-eater of tropical seas.

The thresher shark (*Alopias*) is aptly named because it has an extremely long, whip-like tail, approximately equal to the length of the body. It gives the shark tremendous driving power and it rapidly encircles schools of small fishes. The shark also thrashes the surface of the water with its tail to help frighten the fishes into tighter groups so that they are easier to catch.

The only spotted shark – and the largest fish alive today – is the whale shark (*Rhincodon typus*), which reaches a length of 15 metres. Much has been written

Above: *Only 12 species of sharks out of some 200 species rate as dangerous to man. The rows of teeth are for tearing flesh. As the front rows of teeth are worn down the rear rows move forwards.*

about this monster but surprisingly little is known about its behaviour. Although of monster size, it is extremely docile, according to reports of skin divers who have swum around one of them without injury. It feeds on very small food items, ranging from little fishes to squids and shrimps. Although it has over 300 rows (only 10 to 15 rows function at any time) of teeth in each jaw, this shark strains the food by means of a network of gill rakers that extend from the base of the gill

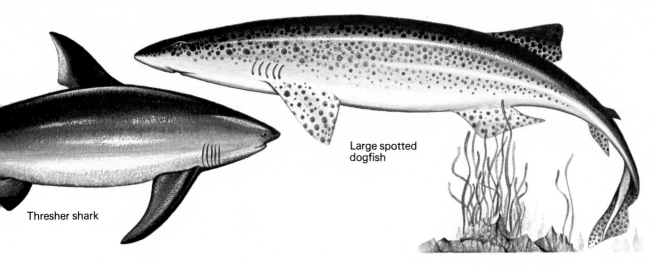

Large spotted
dogfish

Thresher shark

arches into the throat. These sharks are recorded from all the tropical waters of the world, but seem to be most abundant in Caribbean waters and the Gulf of California.

The giant basking shark (*Cetorhinus maximus*) of the North Atlantic is the world's second largest fish, occasionally reaching 14 metres or more. It is like a mackerel shark, but has dropped its carnivorous feeding habit and substituted a diet of plankton, filtering the food in a similar way to the whale shark. The name basking shark is appropriate, because it spends much of its time just floating at the surface or cruising along slowly with its dorsal fin breaking the sea's surface. Recent research suggests that the gill rakers are shed during the winter months and new rakers grow to replace them by early spring.

The hammerhead shark (*Sphryna zygaena*) is one of the most readily identifiable sharks. The eyes and nostrils are found at each end of the 'hammer'. The distance between the eyes of a hammerhead 5 metres long may be as much as 1 metre. The main diet of these sharks appears to be the related stingrays, as well as other fishes. They can devour the rays whole, and appear to have some kind of immunity to the sting.

The Australian carpet sharks are also called wobbegongs. They are not streamlined for fast swimming, but are richly camouflaged to enable them to hide on the bottom of the coral reefs or on rocks. Resting here during the day a wobbegong looks like a rock overgrown with seaweed. At night, from this position, it will pounce on some unsuspecting victim, or will forage around the seabed.

Some of the most beautiful sharks are the catsharks, which have striking patterns of stripes, bars and mottling. The catshark family includes the common European spotted dogfishes (*Scyliorhinus* species). Although shark-like in appearance, they do not grow to much more than 60 centimetres in length. There are about 30 species of dogfish ranging from shallow coastal waters to deep off-shore waters. They feed on virtually anything that they find.

Left: *A white-tip reef shark swims leisurely over a coral reef searching for food. If smells or vibrations indicate that possible food is near, the shark curves and circles until contact is made. Sight is more important than smell.*

Left: *Profile of a manta ray. The greatly enlarged pectoral fins move up and down so the fish 'flies' through the water. It can leap several metres out of the sea. Although they look rather fearsome they actually filter-feed on small crustaceans.*

do not live on the bottom, breathe like normal fishes.

The guitarfish (*Rhinobatis*) of tropical and temperate waters looks half-shark and half-ray in shape. The elongated body is flattened along the sides of the head and trunk, with the pectorals slightly enlarged like rays. It is classed as a ray mainly because the gills are on the underside of the pectoral fins. These fishes are about 2 metres long, and often travel in large schools, in tropical and temperate waters around the world.

One of the most peculiar rays is the sawfish (*Pristis*). It is found in tropical and brackish waters of the world and is known to enter fresh water. Lake Nicaragua has a well-established population that is now probably land-locked. The saw has between 16 and 35 pairs of teeth along its length, depending on the species. This tool is used not only in digging up sandy bottoms to find food, but

Skates and rays

(ORDER BATOIDEA)

Skates and rays are closely related to sharks, but they are very much flattened, the greatly enlarged pectoral fins being joined on each side to the head, so that together they form a disc. The eyes and spiracles are on top of the head, with the nostrils, mouth and gills underneath. There is no anal fin and the tail is usually slender and often equipped with a sting.

To breathe, these fishes do not take in water through the mouth, since, being bottom-dwelling, too much mud and detritus would be included. Instead, water is taken in by the dorsally placed spiracles and then passes out through the ventral gill chamber's slits. Those rays such as the manta rays, that

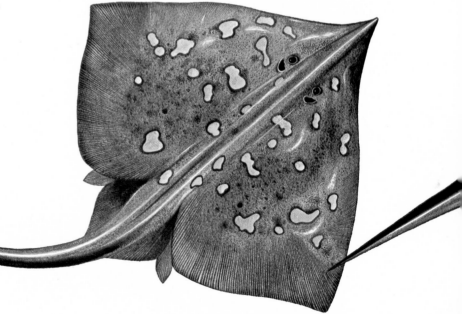

Skate
(*Raja batis*)

also to wound and impale fishes. After catching as many victims as possible, the fish can cruise about eating the dying prey at leisure.

Skates are mostly flat-bodied, and the majority of the 100 species or so belong to the genus *Raja*. One of the commonest skates along the American Atlantic is the little skate or hedgehog skate (*R. erinacea*), which reaches only 50 centimetres or so in length. The largest skate is the American Pacific big skate (*R. binoculata*), which reaches a length of 2.5 metres.

Fortunately only 2 families of rays have venomous spines, but over 100 species are involved. These are the stingrays and whip-rays of the family Dasyatidae and the eagle rays of the family Myliobatidae. Some of these rays reach almost 5 metres and over 350 kilogrammes, and the stinger on the whip-like tail becomes a formidable weapon.

The devil or manta rays have a fearful reputation that is little deserved, although they are extremely large and powerful. One manta was measured at over 7 metres across the wing tips and it probably weighed more than 1,560 kilogrammes. They must be treated with respect but are usually docile by nature. They feed at the surface on small crustaceans that they scoop into the huge mouth.

Chimaeroids

(ORDER CHIMAERIFORMES)
The chimaeroids are sometimes called ratfishes, ghost sharks or elephant fishes. The 25 species have characteristics that in some respects appear to be intermediate between those of the sharks and bony fishes. Shark-like characteristics include the cartilaginous skeleton, the paired claspers of the male and the eggs being laid in horny capsules.

Bony fishes' characteristics include the gills being covered by skin and separate openings for the digestive system and the reproductive system. The males also have a small pair of claspers, which are probably used in courtship, in front of the eyes, but the actual method of using them still remains a mystery.

The short-nosed chimaeras or ratfishes have prominent mucus canals on the head, a venomous spine on the back and a long, rodent-like tail that is responsible for the name 'ratfish'. Ratfishes contain liver oil that is highly prized in the oiling of precision equipment. The most common European species is the ratfish *Chimaera montrosa*, which often reaches 1.2 metres. The smaller American Pacific ratfish (*Hydrolagus colliei*) reaches just over half this length.

The long-nosed ratfishes have a nose shaped like the Concorde jet. They inhabit the depths at between 700 and 2,500 metres and are not often seen.

The elephant chimaeras (*Callorhinchus* species) are so-named because of their spectacular proboscis. They live only in the temperate, and even colder, waters of the southern hemisphere, occurring down to about 200 metres.

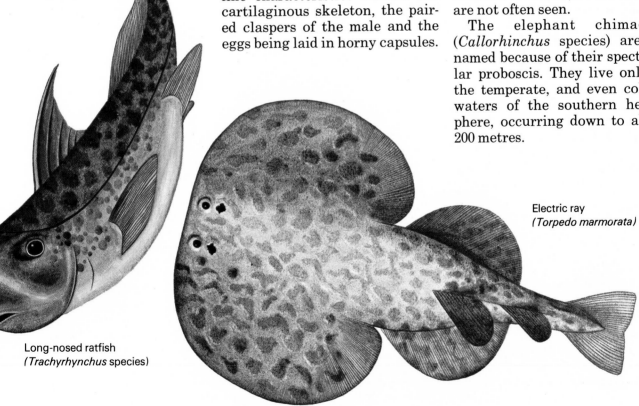

Electric ray
(*Torpedo marmorata*)

Long-nosed ratfish
(*Trachyrhynchus* species)

Bony fishes

(CLASS OSTEICHTHYES)

Ray-finned fishes – primitive survivors

The ray-finned fishes are a very old group and their skeletons are still almost entirely cartilaginous, although the fins are strengthened by jointed bony rays.

The bichirs (ORDER POLYPTERIFORMES) of African fresh waters form a group of some 12 species. They have scales covered with a thick layer of enamel-like substance called ganoin, and have air-breathing lungs. Their generic name, *Polypterus*, means 'many fins' and refers to the 5 to 18 flag-like finlets that can be raised along the top of the back. Apart from the fact that they eat worms, insect larvae and small fishes, little is known about their habits.

The sturgeons (ORDER ACIPENSERIFORMES) are shark-like fishes that are found in temperate and Arctic waters. Their gill structure and the presence of an air bladder link them to the bony fishes. They are sluggish creatures with a long, bony snout which is equipped with barbels or 'whiskers'. These probe in the mud for small invertebrates, which are sucked up by the toothless mouth.

The royal sturgeon (*Huso huso*), which grows up to 8.5 metres, is found in the Caspian, Adriatic and Black seas. It can weigh up to about a tonne and it is this fish whose eggs or roe supply the highly prized delicacy called caviar. Today large specimens are very rare.

In the Mississippi Valley of America lives a curious sturgeon, called the paddlefish (*Polyodon spathula*). It is also given the name spoonbill or shovel-fish because of the shape of its snout. This is expanded to form a large, paddle-shaped blade, nearly as large as the body. A related paddlefish lives in the Yangtse River Valley in China.

Bowfins

(ORDER AMIIFORMES)

The bowfins belong to another ancient order which used to be widely distributed in the fresh waters of both the United States and Europe. Today a single species of the order Amiiformes survives in the rivers and swamps of the eastern United States. The bowfin (*Amia calva*) has a long, slightly arching, spineless dorsal fin. The male makes a nest in weedy areas and guards the nest after the female has laid her eggs.

Below: *Bubbles rise from an ornate bichir. This African tropical river-living fish can survive for several hours out of water due to the fact that it has air-breathing lungs. Flag-like finlets replace the dorsal fin.*

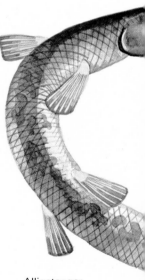

Alligator gar
(*Lepisosteus tristoechus*)

He also chaperones the youngsters for a short time after they hatch.

Garpikes

(ORDER LEPISOSTEIFORMES)
The 7 species of garpike of the order Lepisosteiformes are found only in North America. Some-times they reach over 3 metres in length. They are recognized by their long snouts, which they use to slash sideways at food. The fish prey is effectively pierced and held captive by the long, sharp, needle-like teeth. The most widely distributed is the long-nosed garpike (*Lepisosteus osseus*), found in North American fresh

Above: *The paddlefish, a relative of the sturgeons, is found only in the river Mississippi and its tributaries in North America. The long paddle-shaped snout stirs up muddy river beds. The huge mouth is then opened and minute animals are filtered from the muddy water by gill rakers.*

waters from the Mississippi basin eastwards.

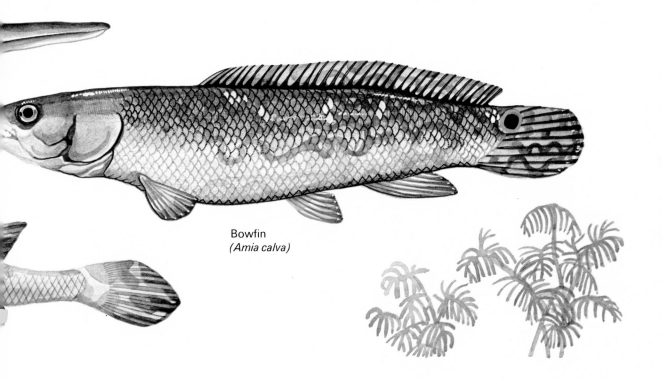

Bowfin
(*Amia calva*)

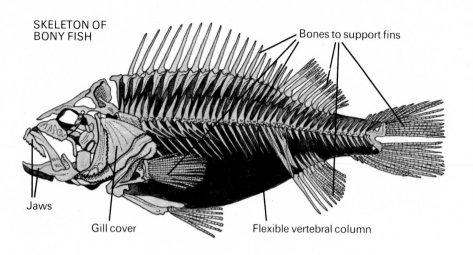

SKELETON OF
BONY FISH

Bones to support fins

Jaws

Gill cover

Flexible vertebral column

Teleosts

(INFRA-CLASS TELEOSTEI)
The teleosts are the most numerous of living vertebrates and evolved over 190 million years ago. They vary greatly in shape, with the bony skull and vertebrae well developed. Their skins are mostly covered with round scales, although a few are naked.

Forming one primitive group are some 12 species of tarpons, large sized (1 to 1.5 metres) relatives of the herrings. They are favourite game fishes along the coasts of Florida and Carolina. They are related to the freshwater mooneyes and the milkfishes. Their large scales are used in ornamental work.

Herrings and anchovies

(ORDER CLUPEIFORMES)
The herring family contains many very important food fishes. Most of the 350 or so species are characterized by their typical 'fish-shape', and by silvery scales that come off the body easily. They live in large shoals near the surface of the sea and may be caught in great numbers. The Atlantic herring (*Clupea harengus*) is found in great shoals, often numbering millions of individuals. The shoals make unpredictable seasonal migrations depending on the temperature of the water, food availability and spawning areas.

The alwife (*Pomolobus pseudoharengus*), a fish some 25 centimetres long, and the shads (*Alosa* species) spend most of their time at sea but enter fresh water to spawn. Their eggs, as in all the members of the herring family, are demersal. This means that they lie on the bottom among gravel and rocks, rather than floating in the surface waters, as occurs in most other marine fishes. Other families in the order are the sardines (the young pilchards), sprats and anchovies.

Eels

(ORDER ANGUILLIFORMES)
Despite their snake-like appearance, eels are true bony fishes, differing from most others in the fact that their scales are very much reduced and buried in the skin, and their paired fins are often quite small. Also there is no dorsal fin and the small jaws are equipped with very sharp, small teeth.

Most of the 300 species are marine, but one family, the Anguillidae, contains the European eel (*Anguilla anguilla*) and the American eel (*A. rostrata*)

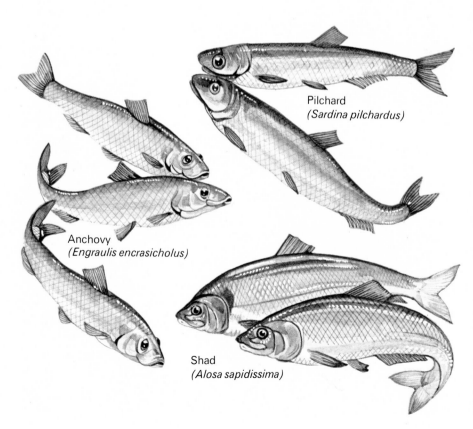

Pilchard
(*Sardina pilchardus*)

Anchovy
(*Engraulis encrasicholus*)

Shad
(*Alosa sapidissima*)

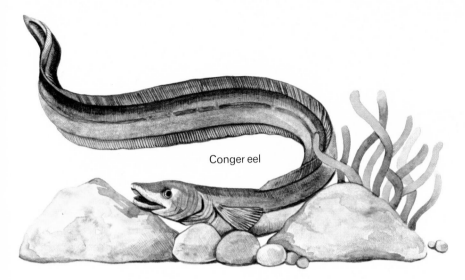

Conger eel

ately coloured and patterned. Most are about 1.5 metres long but some giants have reached over 3 metres. The brilliant banded morays of the genus *Echidna* have flat teeth to grind up molluscs, sand dollars and sea urchins. Like the majority of morays, they hide in narrow crevices between rocks during the daytime and come out at night to forage.

In many countries eels are highly prized as food, some being sold as fresh fish, and others being preserved by smoking and pickling.

whose journeys from fresh waters to the Sargasso Sea area of the Atlantic, to lay their eggs, provide a fascinating story of migration.

Conger eels, are mainly shallow-water-dwelling, but one species, *Promyllantor purpureus*, is found at depths of more than 1.5 kilometres. The moray eels are very large eels found mainly in tropical waters. Many are elabor-

Below: *A moray eel menaces the photographer from its safe hole in the coral. At night this fearsome predator emerges to forage around for food such as small fishes, shrimps and other crustaceans.*

Bony tongues

(ORDER OSTEOGLOSSIFORMES)

These fishes look rather like gigantic pike and live in the rivers of the tropics. They have thick ornamented scales and tooth-like tongues which are used in biting their prey of smaller fishes. The arapaima (*Arapaima gigas*) grows to at least 2 metres and some people say that they reach twice this length in their Amazonian home. Although this is often described as the largest freshwater fish, the South American catfish may be as large or larger. The arapaima breeds in shallow water with a sandy bed where it hollows out a nest with the paired fins.

The related arawana of South America and other bony-tongue members of the family incubate the eggs in the mouth or throat until the youngsters hatch.

Elephant-nosed mormyrids

(ORDER MORMYRIFORMES)

These fishes are closely-related to the bony tongues, but are only found in Africa. They have a unique and strange appearance due to the enormous elongated snout which is found in most of the 150 species. Most of them are also capable of emitting micro voltages. They use these currents to find their way and detect other fishes, enemies, and food such as worms, insect larvae and other invertebrates.

Above: *The impressive arapaima, a very large freshwater fish of South American rivers. It usually grows to over 2 metres and some are said to have reached twice this length with a weight of 200 kilograms.*

Elephant-nosed mormyrid

Fish migrations

After birds, fishes are the greatest migrants of the animal kingdom. Most fishes are constantly on the move, looking for food or merely being carried along by the currents. However, many are true migrants and quite a lot of information has been gathered about the movements and journeys of fishes. Many details have been revealed by tagging fishes with various kinds of labels.

Salmon are renowned for their migrations. There are several species that migrate, including the Atlantic salmon of European waters, and the humpback, sockeye, red, king and chum salmon of the Pacific coasts and fresh waters. All these salmon have two phases to their life histories, a freshwater phase and salt-water phase. After spending some time at sea, the salmon enter river mouths and journey upstream to spawn in small fast-running streams with gravel bottoms. To reach these spawning grounds the salmon must swim powerfully against the currents, often having to make leaps of 3 to 5 metres to clear waterfalls. After finding a partner,

the female scrapes a trench or redd in the gravel and lays her eggs in batches of about 1,000. The male then discharges his sperm over them. The object of the arduous journey is now achieved and usually the adults drift off downstream in a weak state and very few survive the seaward journey.

What is really amazing about a salmon's journey is that the adult is able to find the actual stream where it was born. The mystery is far from being fully understood, but it is thought that these fishes can navigate by the sun's and moon's rays and that they also use their excellent sense of smell to distinguish the differing odours of one stream from another.

European and American eels (*Anguilla* species) are also famous migrators. Adult eels make their way downriver on the first stage of their long migration to the Sargasso Sea. They spawn here in spring and then probably die. The tiny eel larvae hatch and are known as leptocephalus larvae. These transparent, rather leaf-like creatures were once identified as a separate species. As they develop and grow they drift slowly with the currents of the Gulf Stream. The American larvae reach the east coast within a year, while the European ones take three years or more to reach the coastal waters of Europe. Here they change into thin transparent glass eels or elvers and enter the freshwater rivers. They swim upstream and grow in size and darken in colour. After 4 to 12 years in the rivers they have become mature adult eels and are ready to begin their spawning journey.

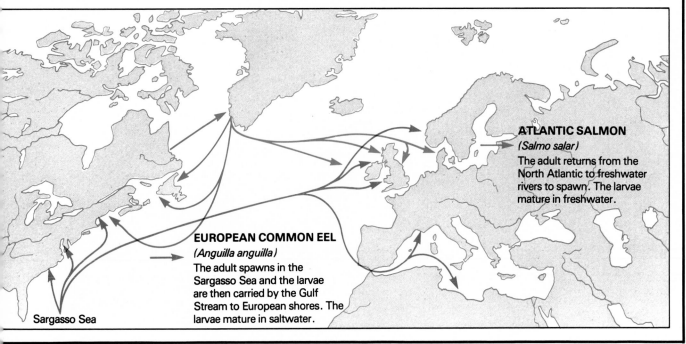

ATLANTIC SALMON
(Salmo salar)
The adult returns from the North Atlantic to freshwater rivers to spawn. The larvae mature in freshwater.

EUROPEAN COMMON EEL
(Anguilla anguilla)
The adult spawns in the Sargasso Sea and the larvae are then carried by the Gulf Stream to European shores. The larvae mature in saltwater.

Sargasso Sea

Light-producing fishes

Certain fishes, from many unrelated families, have the power of producing and emitting light. Those people who have been fortunate to see these fishes, describe the light emission as a vivid, greenish phosphorescent gleam. These fishes live in the pitch darkness of deeper waters of the seas and have evolved this technique for several reasons. The light is used for identifying their own kind, very important when finding a mate. This is possible because each species has the light-producing organs in specific places. Another reason for having these organs is to attract prey to within reach of the mouth. The production of light is due either to the luminous nature of the mucus secreted by the gland cells, the photophores, or to the presence of luminous bacteria.

In the lantern fishes or myctophids the photophores are large and bright, looking like glistening jewels or small mother-of-pearl buttons. Most of the 150 species migrate daily from the depths to the surface waters where they feed on the rich plankton during the night.

In the deep-sea angler fishes the first ray

of the spiny dorsal fin is positioned on the snout and has evolved into a line and bait. The bait is a luminous 'bulb' which attracts smaller fishes to its light; when within range they are gulped into the angler's huge, wide mouth.

Above: *A preserved deep-sea angler displays its fishing technique — a luminous lure at the end of a moveable pole suspended over its huge mouth. Th* *light attracts small fish withln range of the gaping mouth.*

Salmon, pike and relatives

(ORDER SALMONIFORMES)

Salmon and trout are among the best known and most popular fishes among freshwater anglers. All salmon and trout, of which there are some 24 species, have a small adipose (fleshy) fin on the dorsal surface just before the tail fin. They range throughout the northern hemisphere, trout usually being restricted to fresh water, while salmon species spend most of their adult life in the ocean. Salmon return to fresh water to spawn and then usually die from exhaustion and malnutrition. Man has successfully introduced many species into the southern hemisphere.

Trout are smaller than salmon and even when they enter the sea it is thought that they do not swim far away from the estuaries. The European brown trout

Salmo trutta) is the common trout of Europe. The sea trout, once thought to be a distinct species, is now known to be a migratory form of the brown trout. The charr or brook trout (*Salvelinus fontinalis*) is a beautiful species with a mottled pattern, usually with red spots on the sides. Some of the northern forms from eastern North America and Europe do migrate to the sea. The rainbow trout (*Salmo gairdneri*) is another handsome species, with reddish, iridescent bands on the body. In the Great Lakes of America it is called the steelhead.

Pike are voracious carnivores of the fresh waters of the northern hemisphere. The northern pike (*Esox lucius*) is found over most of North America and Eurasia. A large adult can weigh 3 kilogrammes although usually they range between 1 and 5 kilogrammes. They deserve their reputation for ferocity, having been known to deliver a severe bite to the hand of an angler carelessly handling one. They also eat their own kind. They are true cannibals!

In America there are other species of pike. The muskellunge (*E. masquinongy*) of the Great Lakes area can grow to twice the size of a northern pike. The pickerels, as their name suggests, are smaller than the other pike species.

Carp, characins and relatives

(ORDER CYPRINIFORMES)
This order contains the majority of freshwater fishes, many of which are well known. These include the carp (*Cyprinus carpio*), the barbel (*Barbus barbus*) the chub (*Leuciscus cephalus*), the

Left: An adult European brown trout is easily identified by the reddish spots on the sides of the body, each one being surrounded by a light-coloured area.

goldfish (*Carassius auratus*) and the electric eel (*Electrophorus electricus*). One feature that all the members of this order possess is a series of three small bones connecting the swim bladder with the inner ear. This arrangement probably increases the acuteness of their hearing.

The common carp was originally an inhabitant of freshwater rivers and lakes from the Black Sea to Turkestan, but it has been introduced to, and domesticated in, many countries. The Japanese have developed a golden carp that is a spectacular show fish. The mirror carp has a few large scales along its body, while the leather carp has no scales at all.

Minnow is the term often given by many people to any small silvery fish. However, scientifically the minnow (*Phoxinus phoxinus*) of European rivers is a small fish, some 10 centimetres long and with a series of vertical, blackish bars on its sides. There are many species of minnow in America where their common names include shiner (*Notropis*) and fathead (*Pimephales*).

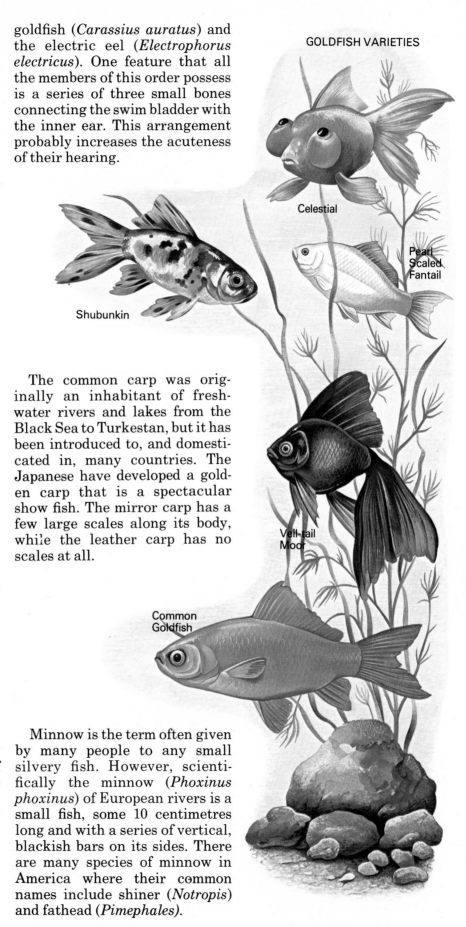

GOLDFISH VARIETIES

Celestial

Pearl Scaled Fantail

Shubunkin

Veil-tail Moor

Common Goldfish

The 500 species of characin are widely distributed in warmer waters from Texas to South America, with a few living in African tropical freshwater. They are minnow- or carp-shaped, but have teeth in the jaws and a small fleshy fin on the back between the dorsal and tail fins. Many of the smaller species are beautifully coloured and have been exported to many countries from South America and Africa, for aquarists. Such names as neon tetra, cardinal tetra, jewel tetra speak for themselves.

The most formidable characin is the piranha. There are about 20 species of this fish, but perhaps only 4 of them deserve the killer reputation. The common piranha (*Serrasalmus nattereri*) is the one most often seen in public aquaria. Piranhas are certainly carnivores but their diet usually consists of smaller fishes, and it is only by accident that they feed on some unfortunate large animal.

They feed in large shoals, sometimes numbering thousands, and if a horse, cow or large rodent such as a capybara, falls under the surface of a piranha-infested South American tropical river, the fishes will snap at it with razor sharp triangular-shaped teeth. Natives often have a finger

Above: *Close-up of the ferocious piranha, the killer of South American rivers. The razor-sharp triangular teeth are usually used on smaller fishes but sometimes huge shoals attack and devour large animals.*

or toe snapped off by the piranhas's jaws.

Catfishes

(ORDER SILURIFORMES)

Most of the 200 or so species of catfish are found in African, Asian and South American tropical freshwaters, but some do occur in northern waters. Most of them have barbels or feelers on the mouth, which are used to find food on the muddy bottom. Catfishes do not have scales, but the armoured catfishes have a bony plate-like armour covering the body. Some catfishes live under ground in the waters of caves or artesian wells. They are usually blind and without skin pigment.

The various naked catfishes show great diversity in form and habit. The largest naked catfish is the European wels catfish

Left: *An African catfish displaying its long feelers or barbels. It uses them to find its way in murky or dark waters. Like all catfishes it has no scales and, also lacking armour, this one is a naked catfish.*

Silurus glanis), which can reach
metres in length. Large ones
will prey on water birds and small
mammals, mainly rodents. The
banjo catfishes of South America
are so named because of their
flattened body and unusually
long tail. In one genus,
Aspredinchthyes, the female
grows a patch of spongy tentacles
on her abdomen, during the
breeding season, and anchors the
eggs to these.

Marine catfishes are mouth-
breeders, the male holding the
eggs and newly hatched young in
the safety of his mouth. These
eggs are sometimes 2 centimetres
in diameter so that the male looks
as if he is holding several huge
marbles in his mouth. During this
time he cannot feed.

Some catfishes of South
America are parasitic on other
fishes, obtaining their food by
eating the gills or drinking the
blood of the host. Another fascinating catfish from Africa is the
upside-down catfish (*Synodontis*).
As you watch this catfish moving
along, it repeatedly reverses the
swimming position from right-
side-up to upside-down and back
again. The electric catfish
(*Malapterusus electricus*) is also
an African species. Large specimens of about 1.2 metres can
produce a shock of up to 400 volts.
The ancient Egyptians considered it a very special fish and it
is featured on certain tomb carv-
ings called pictographs.

Anglers and frogfishes

(ORDER LOPHIIFORMES)

Anglers have flat, squat bodies,
huge heads and wide, cavernous
mouths. There are some 150 species of anglers and related frogfishes living in all types of seas –
warm or cool, shallow or deep.
They are strange fishes that use
the 'rod and lure' technique to
catch other fishes. In the deep-sea

Above: *The yellow frogfish in its brick red colour phase. It takes about a month to change its colour in an aquarium. The stumpy pectoral fins are used as legs to crawl slowly over the sea bed.*

species the lure is luminous. All
anglers are poor swimmers, many
using the paired pectoral fins to
crawl slowly over the sea bottom.
Most are dull coloured with body
outgrowths to camouflage them
against their surroundings.

The largest of the anglers is the
goosefish (*Lophius*), which grows
to a maximum of 1.2 metres. It is
also known as the monkfish and
great fishing frog. In Mediterranean countries it is often eaten
by the human population.

Frogfishes or Sargassum fishes
live in tropical seas and their
peculiar outgrowths and colouring make them look like the seaweeds in which they live.

In certain species of the deep-
sea ceratoid anglers there is a
great difference between the
sexes. The male is a mere dwarf
and lives firmly attached to the
body of a female. The female
provides him with nourishment
by means of a placental-like contact of blood-vessels, but the male
takes in his own water for respiration. The male's body and organs
are poorly developed except for
his reproductive organs. He is
little more than a living bag of
sperm, which the female can draw
on in the breeding season, without
the need for finding and attracting
a mate in the darkness of the deep
seas. The tiny anglers that hatch
can be sexed, as the females show
the beginnings of an angling
device. It is not known when or
how the male seeks out the female.
When mature the largest male is
about 15 centimetres and the
female is about a metre.

Fishes that fly

Blackwing flying fish

Although there are several fishes that are commonly called flying fishes, no fish can truly fly in a bird-like manner; they simply glide through the air on greatly enlarged pectoral fins. The flying fishes of the family Exocoetidae swim very fast, between 25 and 32 kilometres per hour, near the surface of the sea and then break the surface and spread their 'wings'. Flight is undertaken mainly to escape from enem-ies, but it does also occur when a fish is alarmed, at the approach of a ship for instance, and sometimes it occurs for no apparent reason. While gliding through the air the chief motive power of this soaring flight is supplied by the tail, since the pectoral fins are not flapped as are the wings and membranes of birds and bats. It is estimated that a distance of 200 to 400 metres is covered on longer flights. When they land on the decks of boats, some 5 or 6 metres above the surface of the sea, it seems most likely that they have been

Cod and their relatives

(ORDER GADIFORMES)

Cod and cod-like fishes, such as haddock and hake, all have long tapering bodies and they play an important part in the fisheries of the northern hemisphere. The most important species is the Atlantic cod (*Gadus morhua*) which is found on both sides of the Atlantic. These fishes gather in great shoals, especially at spawning time. The European hake differs from the cod species in not having a barbel on the lower lip and it has only 2, not 3, dorsal fins. They are a typically deep-water species with a maximum weight of about 9 kilogrammes. The white hake reaches twice this size.

Other important members of the cod family, are the whiting, the pollack, the ling and various species of rockling. The burbots are cold-water fishes and the only non-marine relatives in the group. They are found on both sides of the North Atlantic, and like cold lakes with stony beds.

Above: *A spotted grouper* (Epinephalus *species*) *chasing another fish. Groupers are related to cod and may grow to a considerable size. The majority of species are carnivorous and have numerous sharp teeth.*

Sticklebacks and seahorses

(ORDER GASTEROSTEIFORMES)

Sticklebacks are small fresh water and marine fishes of the cold or temperate waters of the northern hemisphere. Their common name refers to the stout thorn-like spines which make u carried up on a rising current of air.

Flying gurnards have larger pectoral fins, but their flight is more clumsy and less successful than that of flying fishes. The butterfly fish or chisel-jaw (*Pantodon*) of African rivers and swamps can make short, erratic flights. It can flap its pectorals, but it is not known if this is done when the fish is airborne. The deep-bodied characin fishes from South America are known to beat their pectorals when in the air and against the water's surface during the initial taxiing run.

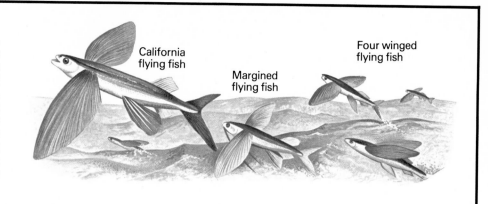

California flying fish

Margined flying fish

Four winged flying fish

the first dorsal fin. The second dorsal fin is more typical. The number of spines which can be raised and lowered usually distinguishes the species. Thus there are the three-spined, the ten-spined and the fifteen-spined sticklebacks. Sticklebacks were the subjects in some of the earliest modern studies on animal behaviour. The three-spined stickleback (*Gasterosteus aculeatus*) of Eurasia and American fresh and brackish waters was studied by the famous zoologist and animal behaviourist Professor Niko Tinbergen. He discovered that many animals are stimulated into action by signs and signals. A male stickleback in courtship-colours of red throat, silvery scales and bright blue eyes, will attack another male because of its red throat. Even a painted red-throated model will be attacked. He also observed the now famous zig-zag courtship dance that the male makes to a silvery female, whose abdomen is swollen with ripe eggs. The male stickleback leads her to the nest that he has constructed by gluing plants together with a sticky secretion from his kidneys. After she has shed her eggs, he immediately enters the nest to fertilize them. He then chases her away and will court another egg-laden, or gravid, female and get her to lay in his nest. When the tiny fishes hatch, he guards them for the first few days.

The related seahorses and pipe-fishes, which are usually less than 30 centimetres long, occur in the warm seas of the world. They are most weird in shape and it is hard to believe that they are fishes. Their small mouth is at the tip of a tube-like snout. This is used to suck up tiny plankton and shrimps. The elongated body is held horizontally in pipefishes but vertically in seahorses. There are no true scales but the body is encased in bony armour. They swim by rapidly rippling the dorsal fin and the paired pectoral fins. The seahorses and pipefishes are remarkable in that the male broods the fertilized eggs in a special brood pouch found on his belly.

Below: *Graceful seahorses are so-called because of the shape of their head. The long tube-like snout has a small mouth at its tip. This has a suction power of pinpoint accuracy to capture small prey that comes within 3 centimetres.*

Scorpion-fishes and gurnards

(ORDER SCORPAENIFORMES)

The fishes of this order are grouped together because all of them have bony plates on the head. Quite a number of them are very spiny, and the most poisonous fishes in the world are found in this large order which contains over 700 species.

Scorpion-fishes are so called because the dorsal, anal and ventral fin spines are venomous. Tropical turkey-fishes or lion-fishes (*Pterois* species) have poison glands in grooves along each of the sides of each spine. The related, ugly, bottom-dwelling stonefish (*Synanceja verrucosa*) has the most deadly of all fish venoms. If a careless swimmer steps on the spines, then he has only about 2 hours to live.

The gurnards or sea robins are small fishes with elongated, platelet-covered bodies. The pectoral fins possess 2 or 3 free rays that function as organs of taste and touch. The gurnard uses these as it 'walks' over the seabed.

Perch and their relatives

(ORDER PERCIFORMES)

The perch order is the largest of the ray-finned fishes, with over

Right: *A dangerous fish with many names, this lion-fish* (Pterois volitans) *carries venom in the spines of its back. It is found in the tropical Pacific where it stripes serve as camouflage.*

6,500 species. They all have spine on their fins, and teeth whe present are small. These marin and freshwater fishes are mainl tropical and the few temperat species are hardly ever found i colder waters. Many fishes in thi group can produce noises an their names indicate this: ther are grunts, croakers and drums.

The European perch (*Perca flu viatilis*) ranges across Europe a far as Siberia. It is usually foun in shoals, feeding on small fishe such as bleak or roach. Ver

Strange births in the fish world

The majority of fish species come together in huge numbers in the breeding season. When a male and female are close together they shed their numerous sperm and eggs into the water. Fertilization takes place by chance and often only one or even none of the young will survive to reach adulthood with the chance of breeding. The number of eggs produced by a female varies. An ocean sunfish can lay over 50,000,000 eggs, while a top-minnow sometimes only lays 1 or 2 eggs.

Some fishes such as sticklebacks and Siamese fighting fishes build nests for the eggs and take care of them once they are laid. Other fishes are live-bearing so that the young get a better start in life, since they are quite well developed at birth, looking like miniature adults. Certain guppies and mollies, as well as many sharks, give birth to well-developed young. The method of reproduction in these fishes is called viviparous.

Some fishes lay their eggs in other animals. The bitterlings (*Rhodeus* species) lay their eggs, through a long tube (ovipositor), into the mantle cavity of freshwater mussels. The eggs develop there, and on hatching the young fry remain inside the protective shells for the first month or so of their development. They feed on the tiny organic matter that the mussel brings inside the shells. A lumpsucker (*Caraproctus*) lays its eggs under the shell of a crab.

A high level of parental care is shown in seahorses and pipefishes. Here the males have brood pouches into which the female lays her eggs. After the young have hatched, for the first few days of their lives, they will still retreat to the pouch when danger threatens.

Above: *A pair of bitterlings get ready to lay their eggs. The female on the left will put her long tube-like ovipositor inside the living freshwater mussel. The male will shed his sperm inside to fertilize them. The young hatch and feed on tiny food particles brought into the shell.*

118

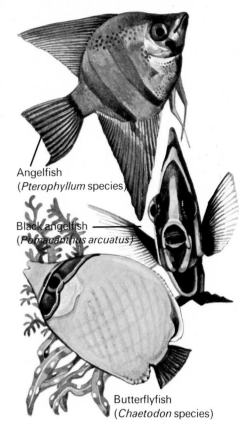

Angelfish
(*Pterophyllum* species)

Black angelfish
(*Pomacanthus arcuatus*)

Butterflyfish
(*Chaetodon* species)

similar to this species is the yellow perch (*P. flavescens*) of North America.

The majority of the beautiful and dazzling tropical coral-reef fishes belong to the perch order. They include the butterfly fishes, angelfishes, cichlids and parrot-fishes. The butterfly fishes with their small, oval-shaped, laterally compressed bodies exhibit a great range of colour and pattern. Thus they are very popular with marine aquarists.

One of the strangest members of the order is the remora (*Echeneis naucrates*) of tropical seas. It has an oval sucking disc on the head with which it attaches itself to sharks, whales, porpoises and turtles. In this way it not only gains protection but is carried about without effort.

Left: *Perch portrait. This freshwater fish of Eurasian waters lurks among the aquatic plants and then suddenly dashes out to snap up smaller fish such as bleak or rudd into its wide gaping mouth.*

119

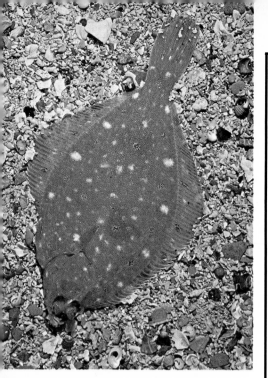

Above: *A plaice, a flatfish species, attempts to camouflage its upper surface by changing its colours and patterns to match with the stone and shell background.*

Flatfishes

(ORDER PLEURONECTIFORMES)
The majority of these fishes, numbering more than 500 species, spend most of their time living on their sides on the ocean bottom. On hatching from surface-floating eggs the young look like typical fish fry. Then after the juvenile stage, one eye moves to the other side of the head, the mouth becomes twisted and the fish begins to settle on the seabed. The fish then always lies on the blind side which remains pale, while the other side is capable of changing colour to match the surroundings.

Usually, all the members of a particular species of flatfish have the eyes and colour on the same side. Thus the halibut, plaice, dab and flounder are right-sided, while turbot and brill are left-sided. The largest species is the halibut, which may grow to a length of over 3 metres and a weight of 275 kilogrammes. The plaice (*Pleuronectes platessa*) is possibly the best known of the

Triggers and puffers

The triggerfishes and pufferfishes are fascinating animals. The triggers receive their name from the locking mechanism of the first and second dorsal fin spines. When the first dorsal spine is raised, the second very small spine moves forward and locks the first into an upright position. A frightened trigger dives into a coral head and erects its spine. It cannot be pulled out, and is completely safe.

The puffers are so called because, when alarmed, they swallow water rapidly and so inflate their body to a balloon shape with all the spines sticking out. They will order and one of the most valuable commercial species of fish landed in European waters.

Lungfishes – more survivors from the past

(ORDER DIPNOI)

Lungfishes can be traced from Devonian times, nearly 400 million years ago. Today there are only two families, containing three genera. The single Australian species, *Neoceratodus forsteri*, retains several primitive characteristics, such as the fleshy-lobed rayed fins, large scales, heavy body and a single air-breathing lung. The South American lungfish (*Lepidosiren paradoxa*) and the African lungfishes (*Protopterus* species) are the most specialized with slimmer bodies, smaller scales, rayless fins and the ability to breathe mainly oxygen from the air, through their paired lungs. The African and South American species are able to withstand droughts by making burrows in the mud and aestivating until the rains come again. The more primitive Australian species dies rapidly if its waters dry up.

The coelacanth

The first living coelacanth seen by scientists was caught near East London, South Africa in December 1938. It was a large fish, 1.5 metres long and was trawled at a depth of about 40 fathoms. Although it was some time before Professor J. B. L. Smith inspected the decaying specimen, he was able to identify it. It was the coelacanth (*Latimeria chalumnae*), a member of a group that had been reputedly extinct for some 70 million years. Further specimens did not come to light until 1952, but since then Professors J. Millot and J. Anthony have examined some 70 specimens and written detailed papers. They have shown that the fish has a very simple heart, even when compared with other fishes. The lobed fins are like those of the fossils and it was surprising to find that they could revolve through 180°. Mature eggs, some 9 centimetres in diameter and few in number (19 were found in one specimen), have been found. As yet it is not known if the female gives birth to live young or lays eggs.

In 1972 a live specimen was caught and for some hours its swimming movements were observed. Although dying and in a small tank, it showed us that the fish moves mainly by sculling its second dorsal fin, the anal fin and the pectorals. There is still a lot of information yet to be obtained about this living fossil.

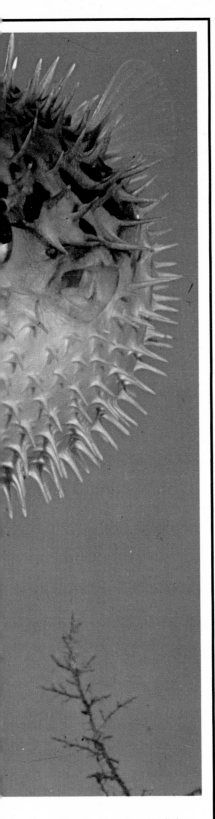

ove: *Normal and inflated pufferfishes.*

o take in air if pulled from the water ey can grow to just under a metre in gth, and so, when blown up, they look te large and if a predator still attempts to allow one it has to deal with the round pe and the spines.

Amphibians (CLASS AMPHIBIA)

The amphibians are the most primitive group of vertebrates living on land. The group includes the frogs, toads, newts and salamanders. Adult amphibians are air-breathing animals, unlike their ancestors, the fishes. Internally, amphibians are interesting because their skeleton has the basic pattern found in reptiles, birds and mammals. Adult amphibians also have lungs, which are found in all the groups descended from them.

Externally, amphibians are not so well adapted to life on land: their skins are not waterproof and so they cannot keep moisture inside their bodies. As a result amphibians must live in habitats that are moist or they will die.

Amphibian eggs are not adapted to land life either. They do not have tough, waterproof shells around them and so they have to be laid in water. The egg does not contain enough food to feed the embryo until it is a miniature adult, and so amphibians hatch at a halfway stage, into tadpoles which live in water. They eat continuously so that they can maintain their growth, gradually changing internally and externally into miniatures of their parents.

The fishes were the only vertebrates in existence until the Devonian period, when the first amphibians evolved from lobe-finned fishes. About 400 million years ago they crawled out of the water on to the land. These first amphibians were nothing like the familiar frogs and salamanders of today. They were larger, to begin with, and they looked rather like badly made crocodiles without scales. The early amphibians retained their tails like the present-day newts and salamanders. Such tailed amphibians may move actively, but the loss of the tail in adult frogs and toads has given

them extra mobility on land.

Scientists believed that the amphibians evolved from fishes which were able to take to the land when the waters of the world began to recede. The planet began to change in the Devonian period, and by the Carboniferous period, the great seas on the face of the planet had given way to swampy forests. There were giant tree ferns growing over the surface of fresh waters. As the swamps began to dry up the fishes that could crawl from one stretch of water to another had a great advantage in survival. Those that could use their swim bladders to breathe air had an even greater advantage. Air-breathing, crawling fishes that were carnivores, and so could find food, had the greatest advantage of all. The lobe-fins changed gradually over millions of years into five-fingered and five-toed limbs: the carnivorous amphibians had arrived.

The Carboniferous period was the amphibian heyday. They lived and thrived alongside the giant dragonflies, in the coal forests. Some were very large, such as *Eogyrinus*, a fish-eater which was 4.5 metres long.

The giant amphibians were still to be found in the next period, the Permian, but the reptiles had evolved by then, and the reign of the amphibians was beginning to decline. *Eryops* was a large Permian amphibian. It was about 2 metres long, and very numerous. Some large amphibians survived until the Triassic period,

Above, left: *A bullfrog's* (Rana catesbeiana) *head showing the moist, scale-less skin used in respiration.*
Above: *Common frogs* (Rana temporaria) *spawning in spring. Most amphibians must return to water to lay their eggs.*

about 200 million years ago. One of the real giants, *Mastodonsaurus*, with its 120 centimetre skull, lived then. However, the close of this period saw the end of the amphibian giants, and only the small amphibians survived until the present day.

There are 3 modern groups of amphibians. They are the legless apodans, the tailless frogs and toads or anurans, and the tailed newts and salamanders or urodeles, which look like small versions of the old giants.

Newts and salamanders

(ORDER URODELA)

Newts and salamanders are the least specialized of all the amphibians, having four legs and a tail. They spend part of their lives in fresh waters, but they leave the water and take to the damp woods, out of the breeding season. The salamanders are better adapted to life on land than the newts.

The life history of the crested newt (*Triturus triturus*) illustrates a typical amphibian life cycle. These newts live in the temperate regions of the world where the winters are cold. During the cold weather, the animals hibernate under logs or stones. They awake in the spring when the weather is warmer and make their way to ponds and calm fresh waters to breed. The male

Permanent children

Some urodeles are very interesting because they never grow up. They spend their entire lives as tadpoles, and can reproduce as such. The axolotl (*Amby-stoma*) is a famous example that lives in freshwater lakes around Mexico City. It remains a tadpole all its life because its surroundings lack iodine. Without iodine, the axolotl cannot complete its development, so that it simply grows into a large tadpole, retaining its feathery gills, and eventually even breeding. It lays eggs which hatch to give small tadpoles, which in their turn grow larger without becoming adults. Most are dark brown, but some are albinos.

Above: *The tiger salamander, the adult form of the axolotl.*

Cave-dwelling olm

Scientists discovered that if the axolotl was injected with iodine, it completed its development and turned into a tiger sala-mander. Tiger salamanders are found all over North and Central America. It is only where the waters lack iodine that they remain as axolotls.

There are several other urodeles that never grow up. In America *Siren* and *Necturus* are found. In Europe, the mysterious olm (*Proteus*) is really a permanent tadpole. Even if it is injected with iodine, it remains a tadpole. It lives in pools in caverns in south-eastern Europe, completely in the dark. Scientists believe that it is a living fossil, and that its ancestors never left the water to live on land. Its strange life, protected from predators, has helped it to survive.

crested newt gives the species its name: during the breeding season he grows a crest along his back. At the same time his belly becomes a bright orange or yellow. He is now ready to look for a female to court, and when he finds one, he dances for her, and if she is ready to mate, she responds. The male then places a small sac, the spermatophore, full of sperm on the bottom of the pond and the female picks it up. Her eggs are fertilized inside her body, after which she glues single eggs on to water-plant leaves.

The eggs hatch into tadpoles with 3 pairs of feathery external gills. At first they clamp their jaws to plant leaves but then they become carnivorous and swim about. After a while their gills shrink and their legs appear, their mouths grow larger and eyes develop. Inside them, their lungs have developed and they can be seen swimming to the surface and gulping air. They are now miniature adults. The newts remain in the water until the autumn. As the weather becomes colder, they climb on to the land and find a suitable spot in which to hibernate until the spring.

Salamanders' lives are similar to those of newts, except that they usually live in warmer areas, and so do not need to hibernate. The European fire salamander (*Salamandra maculosa*) is well adapted to life on land. It spends its time in woods, sheltering under logs. Its reputation for fire-dwelling resulted from this habit. When logs in which salamanders were sheltering were put on to a fire, the animal understandably ran out, giving rise to the belief that salamanders were born in fire!

Above, right: *Great crested newts in breeding colours. The male sports a fine crest and orange belly, the female being crestless and less colourful.*
Right: *The European fire salamander is clumsy in water and liable to drown, so it spawns in shallow water.*

Blindworms

(ORDER APODA)

Blindworms, or caecilians, are highly specialized amphibians, which have become adapted to a life underground. This little-known group of over 150 species lives in damp soil in tropical regions. They are legless with strange, ringed bodies, and really do look like giant earthworms. Their underground life means that they have no use for eyes, and so they have lost them during the course of evolution.

Caecilians have developed new sense organs for their underground life. On each side of their heads are tentacles which are stretched out to touch and smell the earth around the animal. These help the blindworms to catch their food of slugs, earthworms and any other small animal that they find in the soil.

Adult blindworms do not go near water. In fact, if they are put into water, they will drown. The

Above: *A blindworm is also known as a caecilian. Although it looks rather like a common earthworm, this animal is a highly specialized amphibian. It has lost its legs and its eyes during its evolution to become adapted to living underground.*

only time that they are seen above ground is when flooding waterlogs their underground homes. Blind worm tadpoles do live in water, however. The female lays her eggs in a burrow in a bank, very near to a pool. After laying her eggs, she often coils round them to protect them. The eggs have to be fertilized inside the female's body, because mating takes place out of the water. In some cases the eggs hatch inside the female's body, so that she gives birth to live tadpoles. Otherwise, the tadpoles wriggle through the loose, moist soil to the water. They drop into it and lead a fairly typical tadpole life, except that they do not have external gills, just a breathing pore. This pore seals up when their lungs have developed, and they crawl out of the water to burrow into the soil.

Frogs and toads

(ORDER ANURA)

The frogs and toads are the most highly specialized amphibians of all, but they are also the most adaptable. Although most of them return to water to breed, they are better adapted to land life than the rest of the modern amphibians. Frogs and toads live in temperate woods, in tropical forests, in grasslands and even in deserts. There are frogs that live in trees and toads that live underground. They do not live in very cold regions because they are cold-blooded, like fishes, and cannot survive extreme temperatures. There are no true marine amphibians either, although there is a toad named *Bufo marinus*, but it actually lives in woods. The natterjack toad (*Bufo*

Right: *Leaping frog. The long hind legs are specially adapted to push frogs from the ground. The shorter front limbs are muscular and very strong so that they can withstand the shock of landing.*

FROG LIFE CYCLE

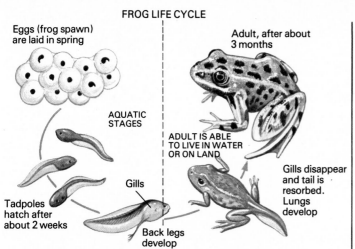

Eggs (frog spawn) are laid in spring

AQUATIC STAGES

Adult, after about 3 months

ADULT IS ABLE TO LIVE IN WATER OR ON LAND

Gills disappear and tail is resorbed. Lungs develop

Tadpoles hatch after about 2 weeks

Gills

Back legs develop

NEWT LIFE CYCLE

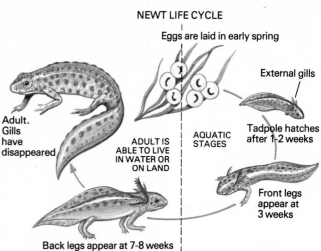

Eggs are laid in early spring

External gills

Adult. Gills have disappeared

ADULT IS ABLE TO LIVE IN WATER OR ON LAND

AQUATIC STAGES

Tadpole hatches after 1-2 weeks

Front legs appear at 3 weeks

Back legs appear at 7-8 weeks

calamitus) occasionally swims in brackish waters, but it cannot be described as marine.

The anuran skeleton is specially adapted for jumping. Frogs are better jumpers than toads, but both have lost their tails and have long back legs designed to push them off the ground. The front legs are very strong, to withstand the shock of innumerable landings. These specializations have evolved over more than 300 million years. The remains of the ancestors of today's frogs and toads can be found in Carboniferous rocks.

Both frogs and toads normally have to return to water to breed: the eggs are fertilized externally, and the sperm need water to swim to the eggs. Their life cycles are very much like that of the newt.

Toads are more likely to move from place to place than are frogs. They are better adapted to drier conditions, with thicker, wartier skins than frogs. Some toads actually live in deserts, burrowing into the ground to preserve their moisture. Other toads burrow out of trouble. The burrowing toad (*Rhinophrynus dorsalis*) digs itself in backwards; the shovel-like back feet help it to disappear from view feet first. If that is not alarming enough, the toad blows itself up like a balloon, to make itself look large and frightening.

Toads have some interesting ways of looking after their eggs. They are often protected by strings of jelly and look like wet knitting wool with knots in it. The male midwife toad (*Alytes obstetricans*) winds his strings of eggs round his back legs and carries them about with him. The male clawed toad (*Pipa*) presses the eggs into the female's back, and she carries the eggs about in little pockets until they hatch.

Both frogs and toads are able to catch quite fast moving prey, although they are not very fast movers themselves. They do this with the aid of the tongue, which is fastened at the front of the mouth, and can be shot out at high speed to hit flying insects. The end of the tongue is sticky, and the helpless insect thus trapped on it is drawn back into the mouth. Both frogs and toads are carnivorous, and will eat anything that they can swallow.

Frogs are more colourful, noisier animals than toads. They are excellent jumpers and can spring away from danger. One African frog is said to have jumped 4.5 metres. Frogs often attract their mates by calling. The bullfrog can produce a noise guaranteed to prevent human sleep at some distance, while the tree peepers tinkle like bells.

The tree frogs are some of the most beautiful and interesting of the amphibians. Most live in the warm, damp tropical rainforests of the world. They have deserted the ground for the trees, and many do not even return to water to breed. The finger and toe tips have become enlarged, to help the frogs to grip when they jump from

Poisonous frogs and toads

Frogs and toads have no sharp teeth or strong claws to protect them from predators. They can escape by burrowing or jumping, but if they are caught, some have a second line of defence. They have glands in their skin which exude poison.

Some of the poisons are fairly harmless. A dog seized a natterjack in its mouth, and was seen to drop it very quickly. It foamed a little at the mouth, but suffered no apparent ill effects after it had washed out its mouth in a puddle.

Other poisons are extremely dangerous. The arrow-poison frogs produce a very virulent venom. These brilliantly coloured frogs live in Central and South America, where the Indians collect the poisons to tip their arrows. The poison is collected by holding the frog over a fire. The poison, released by the animal's sweating, is scraped from its skin into a jar.

The kokoi frog produces the most powerful of all known poisons, more dangerous than any snake venom. It is called batrachotoxin, and less than 1/3,000 gram will kill a man.

Between these two extremes there are poisons of all intensities. For small, soft skinned animals with no visible weapons, the frogs and toads are very well able to defend themselves.

Below: *An arrow-poison frog from South America warns predators of its deadly nature by its bright body colours. Natives collect them and extract the poison carefully so that they can tip their arrows with it for hunting.*

twigs and leaves. They are often beautifully coloured, either as a warning of the poison that they carry or to match their surroundings.

Tree frogs have some strange breeding habits. They have taken to building nests. The Smith frog (*Hyla faber*) builds a beautiful mud nest, with polished walls, in shallow waters. Several frogs stick leaves together and lay their eggs in jelly inside the funnel. These leaves overhang the water, so that the tadpoles can drop into the water when they hatch. Other frogs lay their eggs in nests of foam. The eggs hatch in the foam and the tadpoles develop there, and never live in water.

Above: *The male midwife toad carries the fertilized eggs round his hind legs until they hatch as well-developed tadpoles.*
Right: *A painted reed frog calling. The balloon throat sac is blown out with each sound.*

Reptiles (CLASS REPTILIA)

Rhamphorhynchus

Diplodocus

Brontosaurus

Allosaurus

Brachiosaurus

Reptiles evolved from amphibians about 300 million years ago, in the Carboniferous period. They remained on the planet for 200 million years as the largest and most important vertebrate group. They lived through the Permian, Triassic, Jurassic and Cretaceous periods, at the end of which time most of them died out. A few groups of small reptiles survived longer, and four of those groups exist today.

The reptile skeleton has the same basic pattern as that of the amphibian, with four legs and a tail. Reptiles are better adapted to life on land, however. They have stronger teeth than amphibians, and stronger and more efficient legs, so that they are faster, more formidable hunters. Most important of all, they have evolved waterproof skins, with scales, which keeps the moisture inside their bodies. Reptiles' eggs have a protective, waterproof cover, or shell, and so do not have to be laid in water; quite the reverse, in fact, since the marine turtles come ashore to lay their eggs on dry land. Thus, except for drinking, reptiles are independent of water.

As reptile eggs are surrounded by a tough shell they have to be fertilized inside the female's body, before the shell surrounds the egg and yolk. The fertilized eggs are laid and, after a period of time, hatch into fully developed small versions of the adults. Reptiles need no intermediate stage, like the amphibian tadpole, for the large reptilian egg contains enough yolk to allow the embryo

Above: *A scene which might have occurred in the age of dinosaurs. Although the herbivorous* Brachiosaurus *weighed as much as 7 elephants, it was no match for the ferocious* Allosaurus *which had sharp, tearing teeth and claws.*

to complete its development without an external feeding stage.

Unlike birds and mammals, reptiles continue to grow until they die. As long as there is food available, reptiles grow. Unfortunately, their scales do not. So, at intervals, when the skin becomes too tight, it is shed, a new one having developed beneath. This process is called moulting. If there is plenty to eat, a reptile grows quickly and moults frequently, but if food is scarce, it grows slowly and rarely moults.

The most famous reptiles are not alive today, having died out about 100 million years ago. They are, of course, the dinosaurs. Dinosaurs include some of the largest animals ever to have lived on land. *Diplodocus, Brontosaurus,* and *Brachiosaurus,* were gigantic four-legged herbivores, reaching 25 metres in length. Some dinosaurs were carnivores, *Tyrannosaurus rex* being the best known, and again a giant at over 17 metres. *Allosaurus* was another. These carnivores walked on their hind legs, balanced by their long tails. Some dinosaurs were small, about the size of a chicken, and indeed they were bird-like. These, may in fact, have been the ancestors of the birds.

There are other, less well-known, extinct reptile groups. The pterosaurs were flying reptiles, with tiny legs and great leathery wings. The giant ptero-saur *Pteranodon* had a wing span of nearly 8 metres and the weight of its huge toothless beak was counterbalanced by a backward projection of bone on the head. Plesiosaurs formed another group. They lived in the seas, feeding on fishes. One reptile group, the pelycosaurs, had tall sails on their backs, and it is thought that these reptiles were probably the ancestors of the mammals.

There are four orders of reptiles living today. They are the Chelonia, the turtles and tortoises; the Crocodilia, the crocodiles; the Squamata, the lizards and snakes; and the Rhyncocephalia. This last group is very small and includes only one species.

The tuatara is the only living member of the Rhyncocephalia, its relatives having died out a 100 million years ago. It is a lizard-like animal that survives on some 20 small islands in the Cook Straits, between the North and South islands, and in the Bay of Plenty, of New Zealand. It has survived because it is safe from predators and now, since it is interesting to scientists as a living fossil, it is a protected animal. It lives in a burrow that it has dug or that was originally made by seabirds. In fact it is quite common to find many shearwaters and tuataras sharing the same burrow.

Below: *The tuatara* (Sphenodon punctatus) *of New Zealand. This primitive reptile is often called a living fossil because all its close relatives only exist as fossilized remains in rocks. It was once a source of food for the Maoris and so it became extinct on the mainland.*

Crocodilians

(ORDER CROCODILIA)

The crocodilians form an ancient group that has been on our planet since the Triassic period, over 200 million years ago. They are built on the typical reptilian pattern, but are well adapted for living in water. Crocodilians, which can be described as giant armoured lizards, have survived because they have few competitors; they are the largest, air-breathing carnivores in fresh waters. Their way of life cannot have changed much in the millions of years that they have existed. The crocodilians' extinct ancestors are easily recognizable as such for they, too, have long, narrow snouts and large numbers of pointed teeth. Their prey must have changed, however. *Theriosuchus* was less than 60 centimetres long, and presumably ate mouse-sized mammals. *Phobosuchus hatcheri* was about 13.5 metres long and probably ate fairly large dinosaurs.

The modern survivors of the order include the alligators, caimans, crocodiles and gharials. They live in fresh waters and swamps in the tropical regions of Africa, Asia, Australia and America. Like all reptiles, the crocodilians cannot regulate their own body heat. If the sun is hot, the animal is hot, and since it cannot sweat to cool itself, it will lie instead on a river bank, with its jaws wide open to allow evaporation to take place and heat to be lost in the process. If the weather is cold, so is the crocodilian, as it cannot shiver to warm itself. Neither do the scales keep in heat as fur and feathers do. So the group is forced to live in really warm places, even temperate zones being too cold.

The need to warm or cool itself dictates the pattern of a crocodilian's life. Early in the morning,

Above: *A marsh or mugger crocodile* (Crocodylus palustris) *from India. This opening of the jaws is the way the crocodile cools itself as it is not able to sweat.*

the animals have to leave the water in which they have spent the night and crawl up on to a bank or sand-bar. There they bask in the sun and warm up. The noon sun is too hot for them, however, and so they have to slip back into the water or move into shade to cool off. In the afternoon they bask again and build up warmth for the night. They then return to the water because it does not cool as quickly as does the air at night.

Crocodilians are well adapted to their watery homes. The usual view of a crocodile is a log-like back floating in the water. On looking closer, the head, with the eyes and the nostrils well above the water, can be seen. Both the eyes and the nostrils are raised, so that they remain above water level when the rest of the animal is submerged.

Crocodiles and alligators are very much alike. They have fairly broad jaws and bony plates in rows along their backs. To tell the difference, you have to look at their teeth. An alligator's top teeth overlap its bottom teeth when its mouth is closed, while a crocodile's teeth meet but do not overlap. The fourth tooth in the bottom jaw gives another clue. The alligator's fourth tooth fits into a pit in its top jaw, so that it cannot be seen when the mouth is closed, while the crocodile's fourth tooth fits into a notch on the outside of its top jaw, so that it can be seen. Caimans have bony plates on their bellies, and so they are easy to identify. The false gharial's and the gharial's extremely thin snouts make them easily recognizable, in their turn.

Most crocodilians are fresh-water animals, but there is one exception. The estuarine crocodile (*Crocodylus porosus*) is normally found along and in the mouths of rivers. It lives in nor-

thern Australia and south-eastern Asia, and has also been found in the Solomon Islands and Fiji. It regularly swims between the islands of the Malay Archipelago.

There are only 2 species of alligator, the Chinese alligator (*Alligator sinensis*), living in the Yangtse River basin, and the American alligator (*Alligator mississippiensis*), living in the south-eastern United States. Caimans live in Central and South America while crocodiles live in all the tropical areas of the world.

The crocodilians' life histories are very similar. The male courts the female, often by whirling round in the water and roaring, and then mates with her. Some females make nests of mud or vegetation, or a mixture of both, while others dig a pit. The eggs are laid in batches, of about 15 to 20. They need warmth to hatch, which may be provided by the sun, or by the heat given off by rotting vegetation. In some cases,

Below: *Young caiman. Caimans differ from most other crocodilians in having bony plates beneath the horny scales of both the belly and the back, instead of on the back only. The proportions of the snouts are variable.*

Below: *A nest of a crocodile is unearthed to display the large white eggs. The warmth of the sand heated by the sun helps to keep the eggs at a good temperature for development.*

the hatching crocodilians make noises which stimulate the female to remove some of the debris to allow them to escape.

The young feed on insects and small crustaceans at first, moving on to fishes, frogs and other small animals, as they grow. They grow quickly, at about 25 centimetres a year. The Nile crocodile (*Crocodylus niloticus*) is ready to breed when it is about 2 metres long. It

Left: *Front and side-on views of American crocodile* (Crocodylus acutus) *heads. The large fourth tooth of the lower jaw and the notch in the upper jaw can be seen clearly.*

reaches this length in 5 to 10 years. By then it is trapping birds and larger mammals to eat.

The crocodile lies in wait for animals that come down to the water to drink. It seizes the prey and pulls it into the water. There the animal is stunned by a blow from the crocodile's head or tail, and drowned. The teeth are not suitable for tearing up food, and so the crocodile usually stores its prey for a while, until it rots a little. It will then wedge the carcass firmly, take hold of a leg and shake it until it breaks off. The powerful jaws reduce the food to a swallowable size. Inside the crocodile's stomach there are stones, which it has swallowed, to help it break up its food.

Crocodiles and alligators are hunted for their skins which make beautiful leather. Alligators, particularly, are in danger of extinction because of the past widespread trapping. Al-

thoug[...] a prot[...] ties f[...] unscr[...] it bec[...] to pay[...] Caima[...] plates[...] skins [...] gharia[...] a sacr[...] its range.

The gharial is so unlike the rest of the crocodilians that it has a family to itself. It lives in fresh waters in India. The gharial or gavial (*Gavialis gangeticus*) seems to have a close relative living in Malaya, Borneo and Sumatra, but, in fact, the false gharial (*Tomistoma schlegeli*) is a crocodile.

Gharials are harmless crocodilians. Their very long, slim snouts and not very powerful jaws are well adapted for catching fishes, but nothing larger. The animal sights its prey with the eyes positioned on the sides of the head, and then makes a swift sideways lunge with the jaws.

The female gharial builds a

Turtles, tortoises and terrapins

(ORDER CHELONIA)

The chelonians include th[...] toises, terrapins and turtle[...] group is another ancien[...] having been on Eart[...] [...] ment they were hatched. Some of them bit my fingers before I had time to remove the shells from their bodies. The length of these new-born creatures was 15 to 16 inches, 9 of which belonged to the long and slender tail.' Adult gharials can grow to 6 metres in length. They spend more time in water than the other crocodilians do. They lie with their eyes and nostrils above the water, sinking their eyes below the surface if they are approached, and sinking completely if the approach is too close.

Below: *The extremely long snout of an Indian gavial is an adaptation for catching fishes. The eyes placed on the side of the head see prey and then the jaws sweep sideways trapping fishes on the needle-sharp teeth.*

...e tortoises. This ...t one, ...for about 250 million years. Like the crocodilians, the chelonians have not changed much since they first evolved. The group is a fairly successful one with over 100 species in it. Because they are reptiles, and cannot regulate their body temperature, chelonians usually live in tropical or subtropical zones. However, their range is larger than that of the crocodilians.

The name usually indicates a chelonian's habitat. Americans often call terrapins and tortoises turtles, but strictly speaking, the turtles live in the sea, although there is one freshwater turtle, the Papuan turtle (*Carettochelys insculpta*). The terrapins live in fresh waters and the tortoises, with the exception of the European pond tortoise (*Emys orbicularis*), live on land.

Chelonians cannot be mistaken for any other animal. They live inside protective armour, with their backs protected by a domed carapace and their bellies protected by a flat, bony plastron. Many can withdraw their heads into their shell and block the holes with their limbs – an effective defence.

One disadvantage of this armour is that breathing is difficult. Chelonians cannot expand their ribs to draw air into their lungs, because their ribs are attached to their shells. Instead they have to pump air into their lungs, movements of the head and limbs assisting in this.

Another typical chelonian feature is the lack of teeth. The edges of their jaws are very sharp and bony and are covered by sharp horn. They can eat both animals and plants, and deliver a very nasty bite with these jaws, so their toothlessness is no disadvantage.

There are not many sea turtles left today. The best known are the green turtle (*Chelonia mydas*) of turtle-soup fame; the hawksbill (*Eretmochelys imbricata*), of tortoise-shell fame; the leathery turtle or luth (*Dermochelys co-*

Below: *A young tortoise surveys his world. All the turtles and tortoises have lost their teeth which have been replaced by horny beaks like those of birds.*

riacea); and the loggerhead (*Caretta caretta*). The latter is sometimes found in temperate waters. The luth is the largest living chelonian, growing to 2.5 metres long and weighing 816 kilograms.

Sea turtles come ashore to lay their eggs. The habits of the green turtle illustrate the general pattern. They are migratory animals, leaving their feeding grounds off the Brazilian coast to travel across the Atlantic to Ascension Island to breed – a neat piece of navigation! Mating takes place in the sea at the end of the migration. The female then crawls up the beach, at night, to lay her eggs. She digs a pit, above the high tide mark, by sweeping away the sand with all four flippers. She slowly sinks into the sand until her back is below ground level. Then she digs an egg hole, into which about 100 eggs, each one being the size of a ping-pong ball, are laid.

After she has laid her eggs, the turtle covers them with loose sand. She does not press the sand down, so that it is easy to see where the eggs have been laid. This is unfortunate as men and other predators are fond of turtles' eggs. The luth does pack down the sand and also disturbs a large area, so hiding its nest.

The exhausted green turtle returns to the sea after a night of digging and difficult breathing. Sadly, these turtles are now in danger of extinction, as are the hawksbill turtles. Their breeding grounds are being destroyed and their numbers are dwindling. However, some breeding beaches are now protected and these ancient creatures may yet survive.

Left, above: *A female hawksbill turtle digs a pit in which to lay her eggs.*
Left: *On hatching, the young turtles instinctively head for the sea, and swim off. Many die because seabirds and fish eat them.*

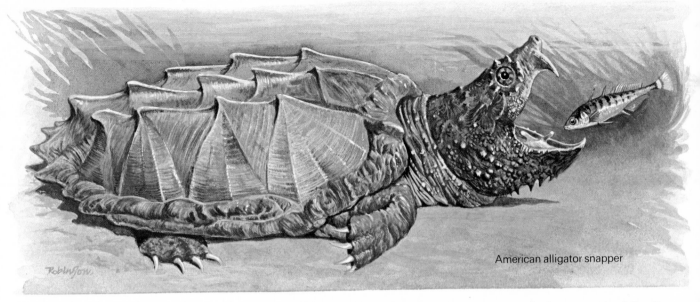
American alligator snapper

The terrapins, which live in fresh water, are found all over the world. Some of them, like the American painted turtle (*Chrysemys picta*), are beautifully marked and coloured. Others, like the Australian snake-necked turtle (*Chelodina longicollis*), are strange looking. The Malayan big-headed turtle (*Platysternon megacephalum*) has a very descriptive Latin name. It means flat-chested bighead! The head is

Below: *Head portrait of a terrapin. Terrapins live in fresh waters in warmer countries of the world. The beautiful shell and head markings identify the species.*

large and covered with horny armour so that it cannot be pulled into the shell.

Some terrapins have a second line of defence. The musk turtle (*Sternotherus*) defends itself by discharging over its attacker a very unpleasant fluid, produced in glands found on each side of the body. Musk turtles lay their eggs under rotting logs or bury them in mud. The hatchlings are quite tiny, only 2.5 centimetres long. Not all terrapins chase their prey. Instead they wait for food to swim into their mouths! The American alligator snapper

(*Macrochelys temmincki*) sits on the bottom of a pond with its mouth open. It has a pink, worm-like tongue, which it wriggles. Fishes, thinking that the tongue is a worm, swim to investigate. They are then quickly snapped up by the terrapin.

The land tortoises are quite a different shape from their aquatic relations. Their carapaces are high and domed. They inhabit tropical and subtropical regions all over the world. Most of them live in dry, sandy places, although some are found in woods. The majority are herbivores, but tortoises have been seen to eat carrion, acting as scavengers.

The Greek tortoise (*Testudo graeca*) and Hermann's tortoise (*T. hermanni*) are the tortoises that are kept as pets all over the world. In temperate places, they need to hibernate in winter. The habit of keeping tortoises for pets has helped us to discover the age to which they live. One famous tortoise was owned by Archbishop Laud. It lived in the grounds of Lambeth Palace, in London, from 1633. There is some doubt about its date of death, one version being 1730, in which case it was over 97 years old, and another being 1753, making it an impressive 120 years when it died.

Giant tortoises

The giant tortoises are interesting, because they have been left behind from another age. They are found on isolated islands in the Pacific and Indian Oceans. They have presumably survived there because there are none of the competitors or enemies, which wiped out their mainland fellows.

The Galapagos tortoises (*Testudo elephantopus*) live, as their name suggests, on the Galapagos Islands. They were described to the scientific world by Charles Darwin, who visited the islands on his famous voyage. The giant tortoise (*Testudo gigantea*), is the tortoise found on several islands in the Indian Ocean. They once lived in the Mascarenes, the Comoros, the Amirantes, the Seychelles and Aldabra, but sadly they are now extinct on many of these islands.

The reason for their extinction is man. These slow-moving herbivores were an easy source of fresh meat for sailors. The animals were so hardy that they actually stayed alive on board, guaranteeing fresh meat for the voyage. The sailors then released any that were left over wherever they happened to end their voyage. In 1759 one ship carried 6,000 tortoises from Rodriguez to Mauritius, an indication of how plentiful they then were.

The giant tortoises reach 1.5 metres in length and 0.75 metre in height. They are herbivores, eating cactus, leaves and berries. They travel about mainly at night, sheltering from the heat during the day.

Above: *A Galapagos giant tortoise eating. First described by Charles Darwin, this reptile can weigh over 200 kilograms, and live 100 years or more. The males are bigger than the females.*

Lizards

(SUB-ORDER LACERTILIA)

The lizards have been on the Earth for about 180 million years, since the Jurassic period, and they are older than their relatives, the snakes.

The lizards are not as conservative as the older groups, the chelonians and crocodilians. Although built on the familiar pattern of four legs and a tail, they have adapted to many different ways of life. There are land lizards, tree lizards, burrowing lizards, aquatic lizards and even an aerial lizard. They range in size from the Komodo dragon (*Varanus komodoensis*), at over 3 metres, to tiny geckos, a few centimetres long.

Lizards cannot live in very cold conditions, but they do manage to live in temperate zones, although the majority inhabit subtropical and tropical lands. They have to hibernate in winter in the cooler areas. Several lizards are able to live in deserts, because as a group, they are well adapted to dry conditions.

Lizards have comparatively long tails. The green lizard (*Lacerta viridis*) for example, is 38 centimetres long, 22 centimetres of the length being tail. Lizards have eyelids, although these may be fixed as a transparent scale, and they have easily identifiable ears. Like snakes, lizards are covered by waterproof scales, and their heads are protected by horny shields.

The life history of the green lizard of southern Eurasia will illustrate a typical life cycle. At the beginning of the breeding

Above, right: *Giant among the lizards, the Komodo dragons are formidable predators and can kill large mammals.*
Right: *The green lizard sunbathes on a rock. This European reptile starts life brown in colour and becomes green.*

140

season, the male becomes brightly coloured, and starts to threaten other lizards. The other males respond with a threat but females do not, so that the sexes are able to recognize each other. Male and female mate and the female lays her eggs shortly afterwards. She digs a nest in the soil and lays the eggs in it. Thereafter she remains near her nest until the eggs hatch. The warmer the weather, the shorter the period before they hatch. They usually hatch 2 to 3 months after laying, the small lizards tearing through the leathery shells with the help of an egg tooth. This is a tooth on the end of the snout, which is shed soon after hatching.

The young lizards feed on insects, spiders, woodlice and other small animals that are available to them. As they grow, they progress to earthworms and small mammals, such as mice, and they will even eat smaller lizards. Adult lizards may eat eggs and fruit.

The thorny devil (*Moloch horridus*) lives in the deserts of Australia. As both its common and scientific names suggest it looks fearsome. It is covered by sharp thorns and seems dangerous, but in fact, it is a harmless animal with weak jaws. It feeds on ants, picking them up individually with its tongue. Unlike the lively green lizard, the thorny devil moves very slowly. If it is frightened, it curls up, rather like a hedgehog.

The thorny devil is an agamid lizard, one of a group of 300 or more species living in the Old World and Australia. It is related to the frilled lizard (*Chlamydosaurus kingii*), which raises a frill around its neck if frightened. The flying lizard (*Draco volans*) is another agamid. It does not, in fact, fly but glides, its 'wings' opening like a fan along its sides and helping it to glide through the air.

Above: *An Australian frilled lizard raises its bright neck flap and opens its mouth wide to scare away any enemy that comes too close.*

Below: *A flying lizard of southeast Asia does not fly but glides on body flaps that act as parachutes when extended from its body.*

Iguanas

The basilisk (*Basiliscus basiliscus*) is an iguanid lizard. The 700 or more species of iguana fill the niches in the New World that agamids fill in the Old World and Australia. The basilisk lives in trees and shrubs near fresh waters in tropical America. When the lizard is disturbed, it takes to the water, where it may sink to the bottom and hide, or it may actually run across the surface. An intruder trying to follow will simply sink!

The marine iguana (*Amblyrhynchus cristatus*) is a relative of the basilisk. It is the only lizard to live in and near the sea. Marine iguanas are strange, greyish-coloured animals which grow to 1.25 metres. The lizards do not spend all their time in the water, but feed on the seaweeds growing on the rocks, following the tide down as it recedes. Some dive into the sea to feed underwater, usually feeding at about 4.5 metres. They usually stay underwater for 15 to 20 minutes, but can remain down for an hour without drowning. The nests are dug into the soft sand on the beach, the eggs being laid at the end of a 60 centimetre tunnel.

Geckos

The acrobats of the lizard world are, without doubt, the geckos. These fairly small lizards are found in all the warmer lands of the world. With their large, round eyes, their colouring and their amazing climbing abilities, they are engaging little animals. Geckos have enlarged, flattened toes crossed by tiny flaps of skin

Below: *Marine iguanas blend in perfectly with the volcanic lava of Galapagos.*

called lamellae. Each lamella carries hundreds of tiny hooks, which catch in irregularities on the surface on which the gecko is

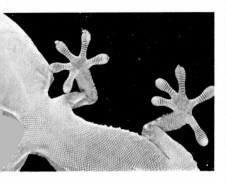

Above: *Hundreds of tiny hooks enable geckos to cling to vertical glass.*

climbing. These lizards can travel across vertical glass, and can run upside-down across ceilings.

Most geckos live in trees or deserts, but some have moved in o live with man, since houses contain numbers of the insects on which they feed. Several species are known as house geckos now. These are usually the smaller lizards, such as the common gecko (*Gecko gecko*). The largest gecko is the tokay (*G. stentor*), which grows to 35 centimetres.

Geckos share an interesting defence mechanism with many other lizards. They can shed the tail at will. If a gecko is attacked it contracts a ring of muscles round its tail and breaks it off. The tail writhes on the ground, attracting the attention of the attacker, while the owner makes a getaway.

Below: *The ghost-like appearance of a hand-footed gecko shedding its skin.*

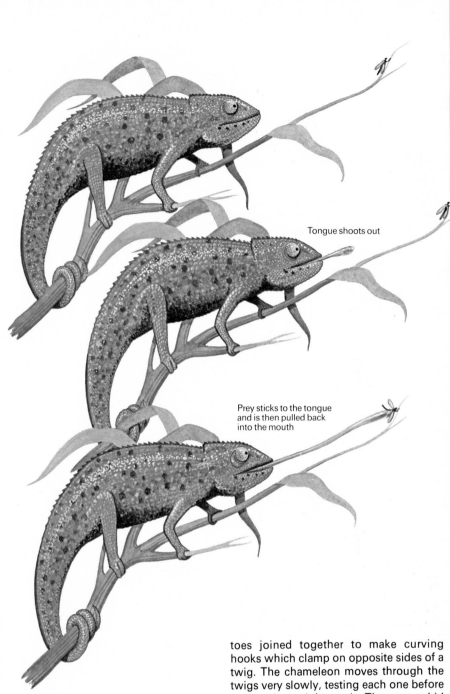

Tongue shoots out

Prey sticks to the tongue and is then pulled back into the mouth

The strange chameleon

The chameleon is a complete contrast to most lizards. It is a grotesque, slow-moving, swivel-eyed animal. There are some 85 species and they are all remarkable animals. They have the strange combination of a slow-motion body and a quick-draw tongue, not to mention their unique, independent eyes, and their ability to change colour.

Chameleons are tree lizards. Their feet are adapted for gripping twigs, with the toes joined together to make curving hooks which clamp on opposite sides of a twig. The chameleon moves through the twigs very slowly, testing each one before putting its weight on it. The eyes whirl round independently of each other, looking up, down, forwards and behind, searching for insects or spiders. When food is sighted, the tongue shoots out, usually hitting the prey, which sticks to the tongue and is pulled back into the mouth. This all happens in no more than 1/25 of a second. Large chameleons will capture birds in this way. Chameleons are masters of colour change. They are able to express their emotion or meet a camouflage requirement by a change in their colour and markings. The skin is sensitive to light intensity and colour change is also controlled to some extent by the eyes.

Skinks

Skinks are streamlined lizards, many of them living in sand, in deserts. Many of the 700 or so species have lines along their bodies, making them look even more elegant. The colours are often as bright and shiny as new paint. The young of some skinks such as *Eumeces skiltonianus* have bright blue tails. They can shed their tails, like the geckos, and the bright colour distracts their predators.

Skinks vary in size from the

Left: *Two heads to confuse any predator is the ploy of this stump-tailed skink. The real head is on the right. This skink has only 2 babies at a time, each one being about half the length of the mother.*

Poisonous lizards

Unlike their close relatives, the snakes, most lizards are not poisonous. Of the 3,000 species of lizards, only 2 produce venom. These are the beaded lizard (*Heloderma suspectum*) and the gila monster (*H. horridus*). Both have venom glands in their lower jaws, between the lips and the gums. The poison oozes from the glands and passes into grooves in the teeth. When the lizard bites its prey, the poison enters the wound from the grooves. This system

is nowhere near as efficient as that of the snakes, and the lizard has to hold its prey in its jaws and chew to work sufficient poison into the wound to have any effect.

The gila monster's poison is not often dangerous to man. Of 34 cases of poisoning, only 8 men died, and those who did were either drunk or unwell! Healthy people survive a bite. The effects are unpleasant, however, including breathing difficulties, dizziness, swelling of the glands and tongue, sickness, fainting and palpitations. The bite area itself swells and

is painful. The venom is a neurotoxin, that attacks the nervous system.

These lizards use the poison mainly as a defence, not to catch their food. They eat eggs, fledglings and young rodents. Beaded lizards and gila monsters are coloured yellow, pink and black, a combination that warns of poison.

Below: *The black and pink scales of this gila monster warns would-be attackers it is poisonous. Venom is stored in mouth glands.*

giant skink (*Corucia sebrata*) at 60 centimetres, to the tiny mabu-yas (*Mabuya* species) which are a few centimetres long. Some of them are rough-skinned. The stump-tailed skink (*Tiliqua rugosa*) has rough scales and a club-shaped tail that looks like a head. The skinks that burrow in the sand are the smoothest skinks. They are called sand-fishes, because they 'swim' through the sand.

Legless lizards

There are several lizards that have lost their legs. This is usu-ally because they are burrowing animals. Legs are not very useful underground, and so most bur-rowing lizards have lost their legs during their evolutionary adapta-tion to living underground.

Some of the skinks are legless. The burrowing skinks of the gen-us *Scelotes* show all the stages from four sturdy legs to none at all. There are complete families that are legless. The worm liz-ards, or amphisbaenids, are one such family. They look like scaly worms and are so much like each other that it needs an expert to identify them. They are burrow-ing lizards, living in South America and Africa. Females lay their eggs in ants' nests where the temperature is warm and steady, and there is a plentiful supply of ants for the hatchlings to eat.

The glass snakes (*Ophisaurus*) and the slow-worm (*Anguis*) are not snakes or worms, but burrow-ing lizards which spend a lot of time on the surface of the ground. They are constantly mistaken for snakes, but on close investigation

Above: *Not a snake nor a worm, but a legless lizard, a slow-worm with its 3 young. If caught as prey by a bird this burrowing animal exudes a foul-smelling liquid which usually causes the bird to drop it.*

their eyes are seen to be the wrong shape for snakes and their tails are too long. About half their length is tail; while snakes' tails take up a third or less of their total length.

Both glass snakes and slow-worms live in open country, on rough ground. The slow-worm spends its time hidden under stones or logs, although it does like to sunbathe. The glass snake hides under fallen leaves or bur-rows just under the surface of loose soil. Both feed on insects, earth-worms, eggs and the occasional fledgling. With no limbs, they cannot tackle faster-moving prey.

145

Left: *A European adder or viper with its young that hatch from their eggs within the body of the mother.*

Snakes

(SUB-ORDER SERPENTES)

Snakes first appeared on this planet about 135 million years ago, in the Cretaceous period. They do not fossilize easily, however, and so their evolutionary history is not well known. Snakes are believed to have evolved from burrowing lizards. Loss of limbs is certainly a feature of a burrowing life, but only a few snakes burrow today.

Snakes evoke a very strong emotional response in humans, and have been both feared and worshipped. Christianity makes the serpent Eve's tempter, the Aztecs worshipped a feathered serpent and the ancient Egyptians believed that snakes were animals from the earliest days of Earth.

One of the reasons for the snake's effect on people is its lack of limbs. It seems to move along by magic. This is because it glides along by means of its scales and by lateral undulations of the body which produce S-shaped curves.

Snakes are able to live in a wider range of temperature than the other reptiles. Most of them live in tropical or subtropical zones, but the adder (*Vipera berus*) is found inside the Arctic Circle in northern Europe.

With no limbs to grasp their prey, snakes either constrict their victims or stun or kill them with poison. They hunt prey down by sight, by sound and by smell. Snakes have no external ears; they pick-up sound vibrations (from the ground) along the length of their bodies. Their large eyes see well, but it is the flickering tongue that usually locates the prey. The tongue is not poisonous, as many people believe. It is picking up minute particles from the surroundings and putting them into the Jacobson's organ. This is a pit in the roof of the mouth, which lizards also have. It helps them to pick up the scent of food and to follow it. Pit vipers have another sense organ that detects their prey's body temperature.

Snakes' difficulties do not end with obtaining food. Mating is not easy when the male has no limbs to grasp the female. Male snakes have therefore evolved double, spined penes to ensure that the sperm enters the female's body and fertilizes her eggs. Some snakes lay their eggs in loose soil or under rotting vegetation. Others hatch their eggs inside their own bodies, apparently giving birth to live young. Wherever they are, the eggs hatch fairly quickly if the weather is hot and slowly in cooler conditions. The hatching snakes tear their way from the eggs with the egg tooth on their snout.

Primitive snakes

Although snakes are not built on the typical reptilian pattern, there is some evidence that they once were. Some of the most primitive snakes still have the remains of back legs. Certain blind snakes and boas still have tiny claws at the base of their tails.

Even the primitive snakes live in a variety of different habitats. The blind snakes (*Typhlops*) and the pencil lead snakes (*Leptotyphlops*) still burrow underground. They are not poisonous. Living underground, they have no use for eyes, and these are greatly reduced. They feed on worms and other small animals that live in the soil.

Constrictors

The boas and pythons are famous for squeezing their prey to death. In fact they wrap themselves round the animal and stop it from breathing; it is very rarely crushed. Boas and pythons cannot kill their prey by poisoning it, and so they have evolved this method. With unspecialized teeth, and no limbs or claws, there is not much else that they could do. Like most snakes they are adapted to swallow prey that is larger than themselves; flexible ligaments and joints allow the two halves of the lower jaw to move apart during swallowing.

Some of the largest snakes belong to this group. The anaconda (*Eunectes murinus*), which lives in tropical America, certainly grows to 9 metres and probably to 11 metres. The reticulate python (*Python reticulatus*), which lives in the Old World, rivals the anaconda in length, growing to 10 metres, while the famous boa constrictor (*Constrictor constrictor*) only reaches 5.5 metres. Some of the boas are beautifully coloured. Sometimes the need for disguise leads to some strange similarities, for instance the Amazonian tree boa (*Corallus caninus*) has almost the same beautiful green and white patterning as the Papuan tree python (*Chondropython viridis*).

Right: *This blindworm is a primitive snake that lives mainly underground.* Below: *A large python has killed a Thomson's gazelle by constricting it. Now it swallows it whole, the lower jaws being adapted to move wide apart.*

Venomous snakes

There are a large number of snakes which, although they are venomous, are not normally dangerous to man. These snakes either lack fangs of any size, or their fangs are situated at the back of the jaws, making it difficult for the snake to sink them into anything that it cannot get right inside its mouth. These snakes outnumber all the dangerous ones in every snake population except the Australian one.

The European grass snake (*Natrix natrix*), which defends itself by squirting evil smelling liquid, and the American hog-nosed snake (*Heterodon platyrhinos*) belong to this group. The hog-nosed snake plays dead, by turning on its back, to confuse predators, determinedly turning back again if it is righted, to prove that it is dead!

Some of these snakes have interesting adaptations to help them eat a particular food. The egg eating snakes (*Dasypeltis* species) have long projections passing from their backbones into their gullets. The snake swallows an egg whole, using the projections to help break it. The contents then pass down into the stomach while the shell is regurgitated. The thirst snakes (*Dipsas* species) have their lower jaws elongated, so that they can thrust them into snail shells and winkle out the soft-bodied molluscs, and eat them.

There are many arboreal snakes, feeding on birds, birds' eggs, tree frogs, squirrels and anything else that they can catch. The African twig snake (*Thelotornis*

Below: *Playing dead by opening its mouth and letting its tongue droop out is the ploy of this grass snake and several other non-poisonous species. Sometimes the snake turns completely on its back to give a better impression of being dead.*

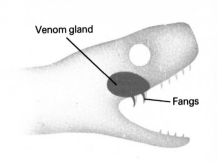
Venom gland

Fangs

Fangs and venom

Snakes inject their venom into their prey by means of their teeth. Some teeth are modified to form fangs. The simplest fangs are found in the back-fanged snakes, whose teeth have a groove in them. The venom is produced by modified salivary glands, and it simply runs down the groove in the teeth. As the groove is open, a lot of venom is lost. The elapids have more efficient tools. Their poison fangs are situated at the front of their mouths, and have deeper grooves, ensuring that more of the venom reaches its destination. The viper fangs are really efficient, as the

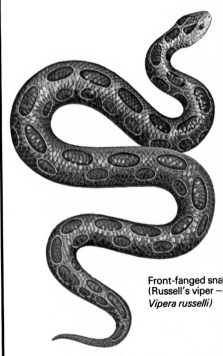
Front-fanged sna
(Russell's viper –
Vipera russelli)

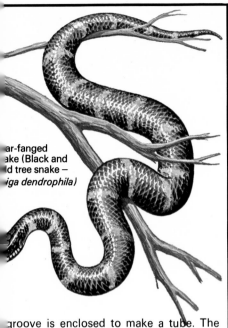

ar-fanged
ake (Black and
d tree snake —
iga dendrophila)

groove is enclosed to make a tube. The venom goes straight through the tooth into the wound, and little, if any, is wasted.

Different snakes have different venoms, although the basic components are similar. There are eight of them. They are: neurotoxins which attack the nervous system; haemorrhagins which destroy blood vessel walls; thrombase, which makes the blood clot; haemolysins which destroy red blood corpuscles; cytolysins which destroy other cells; antifibrins which stop clotting; antibactericidals which kill antibodies; and kinases which begin digestion. Elapid venoms have more neurotoxins, haemolysins and antifibrins, while viperine venoms have more haemorrhagins, thrombase and cytolysins.

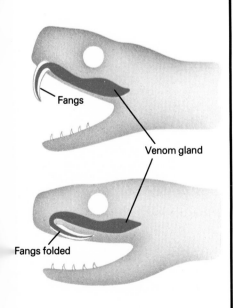

Fangs

Venom gland

Fangs folded

kirtlandii) can hold itself out like a twig, over one third of its body being held out unsupported. It waits in this position for an unsuspecting bird to approach.

Many snakes have taken to the water. The tessellated water snake (*Natrix tessellatus*) and the viperine water snake (*N. maura*) spend over half their lives in fresh waters, feeding on fishes and frogs. They return to land to lay eggs.

A few back-fanged snakes have proved to be dangerous to man. The beautiful green boomslang (*Dispholidus typus*) is known to be poisonous. A famous herpetologist (snake scientist) was handling one in a research laboratory in America, and the

Above, top: *The egg-eating snake can swallow large eggs by being able to distend its jaws and the skin of its neck.*
Above: *A boomslang, a rear-fanged snake, waits poised for a victim such as a bird or chameleon to come within range.*

boomslang managed to bite him. Unfortunately, he died because there was no antivenin for this African snake near enough to save him.

The remaining snakes are dangerous to man. They can be divided into two groups, each with a distinct venom. The elapids have venom in their front fangs, but these fangs are fairly short and rigid. The vipers have very long front fangs, so long that they have to be folded away except when in use.

149

Cobras and mambas

Elapids, of which there are about 200 species, include the cobras, kraits, coral snakes and sea snakes. There are a few of them scattered over Asia, Africa and North America, but by far the largest population is in Australia and the Papuan region, where they equal or outnumber the harmless snakes.

India is home to the kraits and the cobras. The spectacled cobra (*Naja naja*) is the snake used by snake charmers. It looks dramatic because, like all cobras, it spreads the skin of its neck, supported by ribs, into a hood when it is alarmed. It is made to perform by taking off the top of its basket so that light pours in. The cobra rears up, alarmed, opens its hood and sways threateningly. The charmer plays to it, swaying from side to side so that the snake follows the pipe ready to strike. In fact, the charmer is in no danger, as he will probably have pulled out its fangs or even sewn up its mouth! The largest poisonous snake is a cobra, the hamadryad (*Naja hannah*), which lives in India, China and Malaya, and grows to 5 metres or more, as long

Below: *An alarmed Indian cobra rears up and raises its hood. It is very aggressive and will strike rapidly at an enemy or victim.*

as many pythons. It is a snake-eater, and one of the few really aggressive snakes.

The American coral snakes (*Micrurus* species) are brightly coloured, being banded with red, yellow and black. These colours appear throughout the animal kingdom as a warning of venom or unpleasant flavour.

The Australian elapids include the taipan (*Oxyuranus scutellatus*), which is a large snake, growing to 3 metres. The tiger snake (*Notechis scutatus*) and the death adder (*Acantophis antarcticus*) are two more very venomous snakes. The death adder is not a

Above: *This coral snake's warning colours tell of its venom and unpleasant flavour.*

viper, although it looks like one.

There are several aboreal elapids in Africa. The elegant green mamba (*Dendroaspis angusticeps*) and its rather heavier relative, the black mamba (*D. polylepis*) are two famous snakes. There is a tree cobra (*Pseudohaje*) living there too, feeding on birds and arboreal mammals.

The vipers, of which there are about 100 species, are heavier snakes than the elapids, with shorter tails and characteristic triangular heads. They are recognizable as a group, which is not true of the other snakes. The Old World vipers hunt in the same way as other snakes. They include the gorgeously patterned rhinoceros and gaboon vipers (*Bitis* species), which live among the leaf litter in Africa. The mole viper (*Atractaspis*) is a difficult snake to handle, because of its ability to rotate its fangs upwards. If a snake is held by a finger and thumb on each side of its neck and one finger on the top of its head, the handler is usually safe, but not from the mole viper, which can strike upwards.

The New World vipers are call-

Right: A rattlesnake hunts victims by being able to sense the heat of its prey in special heat-sensitive pits on its face.

RATTLESNAKE'S
RATTLE

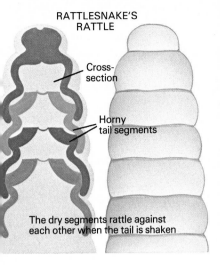

Cross-
section

Horny
tail segments

The dry segments rattle against
each other when the tail is shaken

pit vipers, and they have evolved a unique method of hunting. They have pits on each side of their faces which are heat sensitive. The snake feeds on warm-blooded animals, which it hunts with the aid of its pits. It kills them in the way the Old World vipers do. The fer-de-lance (*Bothrops atrox*) lives in South America, and North America has the copperheads (*Ancistrodon* species) and, of course, the cowboys' nightmare, the rattlesnakes (*Crotalus* species).

Sea snakes – not sea monsters but still dangerous

The sea snakes are the only truly marine reptiles, some of them actually being able to reproduce at sea. They are elapid snakes, with powerful poisons. Their bodies are adapted to their aquatic life, being flattened from side to side to assist the animal as it swims through the water, pushing back in S-shaped bends. The tail is flattened into a paddle shape which steers the snake. They live in warm seas, mostly in the Indian and Pacific Oceans and around Australia.

Some sea snakes come ashore to lay their eggs. *Laticauda semifasciata* once swarmed ashore in its thousands, in the Philippine Islands, to lay its eggs in the sand. Unfortunately, smoked sea snake is a delicacy, and the Japanese fishermen captured vast quantities so that the numbers are now greatly reduced.

The majority of sea snakes do not come ashore to lay eggs. They hatch the eggs inside their bodies and give birth to live young, which can swim away. These snakes are thus totally aquatic. Their nostrils have valves in them to prevent water from getting into their bodies.

The yellow-bellied sea snake (*Pelamis platurus*) ranges from South Africa to Mexico. Its colouring, with a light underside divided from a dark back by a wavy line, is classic pelagic colouring. It makes the animal difficult to see from above and below, protecting it from predators and helping it bear down unseen on its prey of small fishes.

Left: A dangerous sea snake on the seabed. It must rise to the surface every so often to breathe.

Birds (CLASS AVES)

In terms of numbers, birds are more successful than the other warm-blooded animals. There are over 8,000 species of bird in existence today, compared with only 4,200 mammalian species. The success of birds is certainly due to the various adaptations which have equipped them for flight. Their feathers, which have evolved from the scales of their reptilian ancestors, are the key to their success. By evolving an insulating layer of soft feathers, birds could become warm-blooded, and this was a distinct advantage over the cold-blooded reptiles and other lower animals. In some species the feathers have acquired brilliant colours and become important signals in the bird's courtship and social life. Other birds have rather drab plumage to conceal themselves from predators. However, the most important function of the feathers is associated with flight. The strong, long wing and tail feathers provide the flight surfaces, and the body feathers are designed to give the bird a streamlined shape for efficient movement through the air.

Flight requirements have dictated most of the bird's structure and, while the majority of birds are built on a uniform body plan, there is a great variety in size and shape. They range in size from the tiny bee hummingbirds at about 6 centimetres to the tall ostriches at 2 metres or more. These latter birds have lost their ability to fly. In shape, birds range, for example, from the long necked swans and herons to the huge beaked toucans and hornbills and

the short-legged, short-necked, ng-winged swifts.

Birds that fly all have huge eastbone muscles anchored to e keel or sternum of the breast- ne to provide the power for ght. The bones of the body need be as light as possible, and so ey are hollow, although there e fused struts to provide the cessary strength. Teeth, which uld be heavy, are replaced in e birds by horny bills, the apes and sizes of which depend the type of food eaten. Birds do t give birth to live young as ey would not be able to fly with e extra weight of developing ung inside the body. Thus, all ·ds have retained the egg- ·ing habit of reptiles.

Birds evolved over 150 million years ago, the first ones being almost reptilian in structure. The group of reptiles from which birds evolved was probably very closely related to dinosaurs and crocodiles. The oldest fossil bird so far discovered is the well-known *Archaeopteryx*. Its bones were solid like those of reptiles and its short face did not have a beak as advanced birds do, but had jaws with rows of small teeth. It had a bony tail, more like that of a reptile than a modern bird. This feature had been lost in later forms: it obviously disappeared early on in bird evolution. Another reptilian feature is seen in the wings. They were still quite small and definitely similar to the front legs of a bipedal reptile that walked on its hind legs such as the metre-long *Euparkeria*, an assumed forerunner of the birds. Despite all these reptilian features, *Archaeopteryx* definitely was a bird, for it had well developed feathers.

There is a gap of over 60 million years from *Archaeopteryx* to the next fossil birds discovered to date. These include the flightless diving bird, *Hesperornis*, which was about 2 metres long, and the small diving seabird, *Ichthyornis*, which probably lived a similar life to modern-day terns.

The birds we see today began to evolve about 70 million years ago. Fifty million years ago there were herons, gulls and pelicans in existence. Many fossils of these have been found, probably because they lived near water and so were frequently fossilized after death. Flightless bird species were quite numerous at this time, especially on islands and in South America. Madagascar was the home of the elephant bird (*Aepyornis*) which stood over 3 metres high and produced eggs that contained 7.6 litres of liquid. In South America several flightless birds evolved into large carnivorous forms. For example, some 15 million years ago, the 3-metre-tall *Phororhacos* roamed the 'savanna' with its huge, hooked bill, well adapted for killing prey and tearing the flesh, and a skull as large as a horse's.

In the last 50 million years birds have radiated into all the groups identifiable today, and have overtaken mammals in their diversity and distribution.

TYPICAL BODY PLAN OF A BIRD

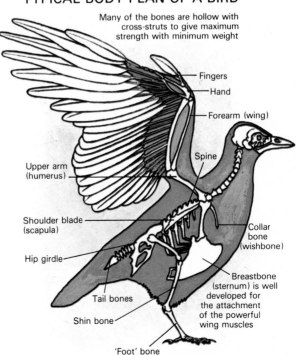

Many of the bones are hollow with cross-struts to give maximum strength with minimum weight

Fingers

Hand

Forearm (wing)

Spine

Upper arm (humerus)

Shoulder blade (scapula)

Collar bone (wishbone)

Hip girdle

Breastbone (sternum) is well developed for the attachment of the powerful wing muscles

Tail bones

Shin bone

'Foot' bone

Flightless birds

Flightless birds are often called ratites, although today it is doubted whether the various orders are closely related. These birds are often quite large, for example the ostriches and rheas, and all of them have very small wings and an unkeeled breastbone.

Kiwis

(ORDER APTERIGIFORMES)

There are 3 living species of kiwi (*Apteryx*), all of which inhabit New Zealand. They are relatives of the recently extinct moa (*Diornis*). Moas reached over 3 metres but the kiwis are the size of domestic chickens, the largest ones not weighing more than 4 kilograms. Kiwis have short muscular legs with strong claws. Their soft, hair-like feathers hide their stubby tails and wings and there are no tail feathers. They

live nocturnal lives in forested areas, and because these have been reduced over the years they are becoming increasingly rare. At night they use their long bills to probe into the ground for invertebrates. The food is detected by their excellent sense of smell – a very rare feature in birds. The nostrils are placed at the tip of the upper beak.

The female kiwi is larger than the male and, in comparison to her size, lays very large eggs. These large, elongated, white eggs weigh about 450 grams. One or two are laid and observa-

Below: *The black and white plumage of a male ostrich distinguishes him from the grey-brown female. Being unable to fly, it relies on speed to escape predators and gives deadly kicks when cornered.*

tions have shown that often it is the male alone who incubates them, a task requiring 75 to 77 days. Little is known about the development of the baby birds.

Ostrich

(ORDER STRUTHIFORMES)

The ostrich (*Struthio camelus*) is famous for being the largest bird living today. There must be few people who could not recognize this bird. An adult male reaches nearly 2.5 metres and weighs up to 150 kilograms. He is identified by his black body plumage, white tail, and plume-like wing feathers. The female is smaller and grey-brown in colour. Both sexes have a long neck and a small head which is covered with short down and bristles, instead of feathers. The thighs are devoid of feathers so that there is less friction when these long legged animals are running. The ostrich has only 2 toes, which correspond to the third and fourth toes of most other birds. The larger of these toes has a flattened nail.

Ostriches used to be quite common over most of Africa and south-west Asia. In the last 200 years they have become extinct in most parts of Africa and are now common only in East Africa. They are found in open habitats, including semi-desert, grassy plains, and open wooded areas. Here they feed mainly on plant matter which includes shrubs, succulents, berries and seeds, but a certain amount of small invertebrates are also eaten. Ostriches are known to swallow pebbles to help grind up the food.

Ostriches live in pairs or in groups of 5 to 15 birds. A male is often accompanied by several females and their immature young. Out of the breeding season large groups of males are sometimes seen. Ostriches are also often seen living with large game mammals, such as gazelles, zebras and wildebeests. They provide a good alarm system for the mammals, since they easily spot predators. In return, they gain a certain amount of protection (although they can defend themselves quite well, usually being able to outrun a lion for instance) and also they pick up insects disturbed by the grazers.

Ostriches breed during or just after the rains. Each male usually has 3 females, which all lay their 6 to 8 eggs in the same nest, a hollow in the ground. The large, creamy white eggs are mainly incubated by the male, but one of the females usually takes a turn during the day. After about 40 days the young chicks hatch. They have black streaks along their bodies, and they can run immediately and so accompany their parents. They are full-grown in about 18 months but do not breed until they are 4 to 5 years old.

Below: *A female ostrich out with her youngsters. They are able to run about from birth, feeding themselves but still running to the protection of the parents if alarmed by anything.*

Rheas – the South American ostriches

(ORDER RHEIFORMES)

The 2 species of rheas live in South America, and they are the largest birds of the New World. The common rhea (*Rhea americana*) stands 1.5 metres high and weighs about 25 kilograms. Although it has larger wings than most other flightless birds it is unable to fly. It prefers bush cover to open ground and is often found near a river or swamp. The common rhea is brown in colour while the smaller Darwin's rhea (*Pterocnemia pennata*) has plumage that is white-spotted on a brownish background.

The male has several females in the breeding season and each hen lays between 11 and 18 eggs in a communal nest. The male usually incubates the eggs for 35 to 40 days. Like their relative, the ostrich, the rheas occasionally mix with grazing mammals, such as bush deer and domestic cattle.

Right: *Colourful portrait of an Australian cassowary. This large flightless bird lives in dense tropical rain forests, rarely being seen in the wild.*
Below: *A South American rhea surveys its environment for any danger.*

Cassowaries and emus

(ORDER CASUARIIFORMES)

Cassowaries and emus are heavily built, flightless birds of the Australasian region. They are covered with a coarse plumage except for the head and neck which are virtually featherless.

There are 3 species of cassowary: the Australian, the one-wattled and Bennett's. They all belong to the genus *Casuarius*. They are found in New Guinea and Australia and a few nearby islands. An identifying feature is the large bony crest on the forehead that probably evolved to fend off obstructions as they run through the jungle undergrowth. The skin of the head and neck is usually highly coloured with blues, purples and reds. They are shy and active mainly at night. If surprised or approached too

osely they will attack and kick
ith their strong legs. A boy from
ew Zealand and several people
om New Guinea have been kil-
d by these birds.

The emu (*Dromiceius novae-
ollandiae*) is about 1.8 metres
ll and is the second tallest bird
the world. It lives on the open
ustralian plains. The sexes are
fficult to distinguish, but in the
reeding season the female grows
lack feathers on her head and
eck, and the small area of bare
in in this region becomes dark
lue. The male's feathers do not
arken and, in comparison to the
male, his blue skin is more
oticeable. The male incubates
e 10 or so eggs for about 8
eeks. The chicks are attrac-
vely patterned with yellow and
rown stripes along their sides
nd back.

ight: *The emu lives on the grasslands of
ustralia, where in the past it was much
rsecuted by farmers.*

Penguins

(ORDER SPENISCIFORMES)

The 17 species of penguin are found only in the southern hemisphere and are better adapted for an aquatic life than any other bird family. Unlike most other birds, all their feathers are of a uniform size and densely cover the entire body surface. The wings are small, narrow flippers and do not fold. The feet are positioned far back on the body and the body skin covers the legs right down to the ankles.

Penguins propel themselves through the water with their flippers, often at speeds of 19 or more kilometres per hour. They use the tail and webbed feet for steering. 'Porpoising' is the usual swimming movement of most penguin species and is used by no other bird. Penguins can also leap several metres from the water to reach, for example, a landing spot on top of a steep ice bank. On land they walk in an upright position, but sometimes they flop on to their belly and 'toboggan' along, propelling themselves with their flippers and feet.

The emperor penguin (*Aptenodytes forsteri*) is the largest penguin, standing 1.2 metres high and weighing 34 kilograms. It breeds on the freezing ice during the bitter cold and darkness of the Antarctic winter. As soon as the female has laid her single egg, she passes it over to the male, who tucks it under the fold of his belly skin, on top of his feet, and so keeps it warm and off the ice. The

Above, right: *A little blue penguin paddling along. The flippers are adapted wings and are used to 'fly' the bird through the water. Penguins are the most primitive group of living birds, having evolved 60 million years ago.*
Right: *Adelie penguins displaying how clumsy penguins are when out of water. They waddle along on webbed feet or toboggan on their bellies on snow or ice.*

Bird feathers

There are several different kinds of feather, but they all have the same basic structure. The primary and secondary flight feathers are very much alike. A flight feather has a central shaft with a series of parallel barbs extending diagonally from each side. The barbs in turn have smaller filaments sticking out from them. These tiny filaments have hooks that lock with other filaments so that a mesh is formed. This provides the firm flight surface that the bird needs when

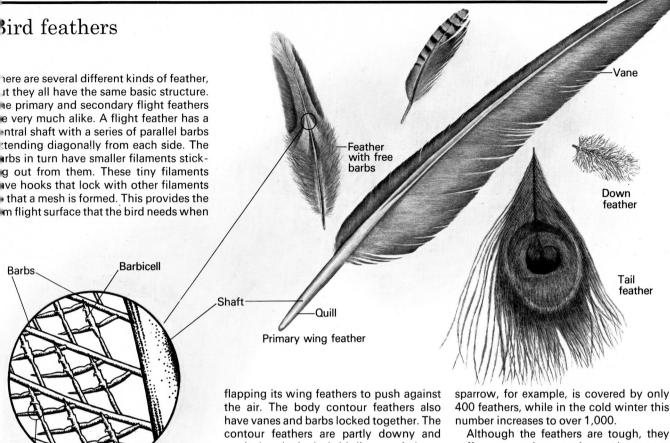

Barbs

Barbicell

Shaft

Quill

Barbule

Feather with free barbs

Primary wing feather

Vane

Down feather

Tail feather

Hook of lower barbule locks in trough of upper barbule

flapping its wing feathers to push against the air. The body contour feathers also have vanes and barbs locked together. The contour feathers are partly downy and partly interlocked. A bird's superb insulation is provided by the down feathers underlying the outer contour feathers. Down feathers do not have vanes.

The number of feathers on each bird varies. Song birds have between 1,100 and 4,600 feathers, depending on the species. The number of feathers on a bird also varies according to the time of year. In summer a

sparrow, for example, is covered by only 400 feathers, while in the cold winter this number increases to over 1,000.

Although the feathers are tough, they suffer wear and tear and sometimes even break. Therefore every grown bird renews its feathers, by moulting and growing new ones, at least once a year. This is usually after the nesting season, and some birds have a partial moult just before the breeding season. This is the time when some birds acquire full finery for courtship and aggressive displays.

female goes off to feed at sea which might be as far away as 80 to 160 kilometres. She is away for about 2 months and the male is left to incubate the egg by himself. The female returns just in time for the hatching of the chick and she feeds it on semi-digested food that she brings up from her crop. The male, now very thin after his fast, goes off to feed. The slightly smaller but similar king penguin (*A. patagonica*) also incubates the egg on the feet but both parents take part in the incubation. These penguins breed on the islands at the fringe of the Antarctic Ocean.

The adelie penguin (*Pygoscelis adeliae*) is the only other penguin

to breed on the Antarctic landmass, choosing patches of bare rocky shoreline. Here the adelies pair and perform an interesting courtship display. The 'ecstatic display' is performed by the unmated male. He stretches his head and neck upwards and flaps his flippers to and fro while uttering a drumbeat cry. Once paired, the birds build a nest of stones in which the female lays 2 eggs. Both parents participate in the incubation which lasts 33 to 38 days. The chicks are born with a grey, downy coat, and when about 4 weeks old the young congregate in groups or crèches of over 100. By 9 weeks they are well grown and ready to depart to the sea.

The little blue or fairy penguin (*Eudyptyla minor*) of New Zealand, is the smallest penguin, at about 36 centimetres high. It nests in burrows or rock crevices. The Galapagos penguin (*Spheniscus meniculus*) lives further north than any other penguin, on the islands of the Galapagos, close to the equator. Although all penguins are basically black and white, several species have attractive head adornments. The macaroni penguin has orange-yellow plumes that project from the forehead and widen and lengthen behind the eye. The erect-crested penguin has a much broader, yellow, curling crest to distinguish it.

Grebes

(ORDER PODICIPEDIFORMES)

The 21 species of grebe are weak-flying birds with long, thin necks and pointed bills. The feet, set far back on the body, have stiff horny flaps on the toes which increase their surface and help the birds when swimming. Most species are satiny grey, or black with white above, and mottled or rufous below. The sexes are alike and their heads usually have crests or tufts. Grebes are found on freshwater lakes all over the world, with a few species moving to the sea coasts in winter. Some of the northern species migrate long distances overland.

Grebes usually fly little and prefer to escape danger by diving and swimming underwater, where they can travel faster than on the surface. They feed under-water on a diet of aquatic invertebrates, some vegetable matter, and fishes.

At the beginning of the breeding season grebes perform elaborate courtship displays. The ritual of a pair of great-crested grebes (*Podiceps cristatus*), for example, includes head shaking, and weed presentation. At the climax of the nuptial dance, the pair tread the water furiously, and then slowly raise their bodies out of the water and meet breast to breast.

The nest is built from rotting aquatic plants and is tethered to growing reeds. The white eggs are covered over if the birds leave the nest so that they are hidden from predators. The young birds can swim and dive as soon as they are hatched. A delightful sight is to see them being carried on the back of a parent with their small striped heads poking out from under the parent's wings.

MUTUAL COURTSHIP DANCE OF GREBES

1. Discovery – one bird dives and surfaces near the other

2. Head shaking – the birds approach each other with heads lowered threateningly, then raise them and shake their bills from side to side

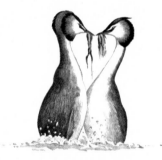

3. Penguin dance – the birds dive and then rise breast to breast with beaks full of weed

4. Inviting – performed near a possible nesting site

Divers and loons

(ORDER GAVIIFORMES)

The 4 species of diver are foot-propelled diving birds, each with a strong pointed bill, small wings and a short tail. Their dense plumage is black and white in colour. The head, neck, and back are spotted, striped, or barred in bold markings, depending on the species.

Below: *A red-throated diver nesting. On land these birds are clumsy and cannot walk properly.*

These Arctic birds are usually seen singly or in pairs but in winter they may form loose flocks at sea. Here they dive for fishes, crustaceans and molluscs, being able to stay underwater for more than 40 seconds.

The red-throated diver (*Gavia stellata*) is the commonest and most widespread member of the family, being found all around the Arctic Circle. It breeds on small lakes and pools in the tundra. A pair will defend their selected breeding site against other divers. Two olive-brown eggs are laid in a loose heap of vegetation near to water. The great northern diver or loon breeds only in Iceland.

Albatrosses, shearwaters and petrels

(ORDER PROCELLARIIFORMES)

The members of this order are oceanic birds, characterized by long tubular nostrils opening out from their hooked bills. This outstanding characteristic has caused them to be named the 'tube-nosed swimmers'. All these birds spend most of their time at sea and come on to land only to nest. They all lay a single white egg, usually in an underground burrow. Only the albatross makes an open nest above ground. All the birds of this order have a distinctive musty body odour which is due to the yellow stomach oil which they discharge from the mouth and nostrils when alarmed. The young are fed on this liquid.

The albatrosses are large, stout-bodied birds with long, narrow wings, which make them superb gliders. The great wandering albatross (*Diomedea exulans*) is the largest of all the world's seabirds. It measures about 3 metres from wing-tip to wing-tip and may weigh up to 11 kilograms. This magnificent bird feeds far out to sea, swooping low over the wave tops and then settling, with wings folded, to dabble for squid and other marine animals which swim near to the surface. It breeds in colonies on the islands of the southern hemisphere. The courtship displays are elaborate and stylized. The egg hatches after 80 days and the chick is cared for by its parents for 8 to 9 months. Consequently, the albatross can only breed every other year.

The shearwaters and fulmars are all migratory and several are among the world's greatest travellers. The slender-billed shear-

water (*Puffinus tenuirostris*), also known as the muttonbird of Australia, migrates north to the Pacific via New Zealand, before returning to the islands in the Bass Straits, between Australia and Tasmania, to breed. These birds were called muttonbirds because the early settlers fed on them. The practice of canning young birds to sell as 'Tasmanian Squab' has been stopped by the Australian Government.

Smallest of the web-footed seabirds are the storm petrels. They range in length from 14 to 25.5 centimetres. Although they look too delicate for ocean life, they are well adapted for it. They are named after St Peter because they can lower their short, webbed feet

Above: *A pair of wandering albatrosses make excellent parents. A single egg is laid on bare or vegetated ground and both sexes share the care of the egg and the chick. The parents probably remain paired for life.*

to skim the ocean surface and thus they appear to walk on the water.

The diving petrels are slightly larger than the storm petrels and as their name suggests they are spectacularly skilful in the water. They dive straight into the sea, for fishes and crustaceans, or to escape predatory seabirds. Their short wings are used in swimming underwater and they can re-emerge directly into flight. Diving petrels live in the southern hemisphere.

Pelicans and their relatives

(ORDER PELECANIFORMES)

This order contains the pelicans, tropic birds, cormorants, shags, darters, gannets, boobies and frigate-birds. All of them have their 4 toes joined by skin to form a web. The black-and-white tropic birds, as their name suggests, live over tropical seas. They are attractive birds with long, streaming, central tail feathers. A distasteful feature, however, is

that the adult birds sometimes kill recently hatched tropic birds for food.

The gannets are heavily built birds with white plumage, dark wing feathers and a creamy head. They live around the shores and islands of northern temperate lands, while their close relatives, the boobies, are tropical in distribution. Boobies are either brown or brown and white in colour. Both gannets and boobies have long, pointed wings and tails, and short, stout legs. They are magnificent divers, plunging from great heights into the water to seek fishes.

The cormorants are medium to large birds, some 50 to 100 centimetres long. Most of the 30 species are black in colour, with the feathers shining with a greenish gloss. These birds sometimes fly inland to freshwater habitats. The common cormorant (*Phalacrocorax carbo*), is found on the coasts of both hemispheres. The Japanese have trained them to catch fishes.

The darters or anhingas are like cormorants but have much longer necks and straight, long, bills. They live in the warm parts of all continents, except Europe. Another name for this bird is snake-bird, referring to the bird's characteristic snake-like way of moving its kinked neck from side to side. The pointed bill is used to spear fishes.

The 8 species of pelican are fish-eating water-birds and are known all over the warmer parts of the world. The pelican's pouch does hold 'more than its belly can' as the old poem states. In fact, the large beak-pouch holds two to three times as much as the stom-

ach. The pouch is used to scoop fishes from the water. In the air pelicans are very graceful and can glide for some distance, but on land they are most amusing as they wobble along on their webbed feet.

The large white pelicans (*Pelecanus erythonocrotalus*) of Eurasia and Africa often fish in groups. Several birds form a line and drive the fishes before them, all the birds dipping their pouches into the water at the same time. The brown pelicans (*P. occidentalis*) of American shores dive into the sea to fish.

Most attractive members of this order are the frigate-birds, which may soar for hours on their long, thin wings. They are also called 'man-o'-war-birds' because they attack other seabirds, to make them disgorge their food. During courtship and nest-building the male inflates his orange throat pouch, which then turns red and stays puffed up for hours. The white-throated female lays a single egg in a frail twig nest that has been built by the male.

Below: *White pelicans feed together by encircling fish and dipping their bills into the water at the same moment.*
Bottom: *A male frigate bird with his colourful throat sac inflated. It can stay puffed up for hours.*

Above, left: *Red-footed booby* (Sula sula) *sitting by its untidy nest. It is unique among boobies in that it nests above the ground away from predatory birds.*
Left: *Gannet greeting. Although to the human eye these birds look identical, the male can recognize a female.*

Left: *Beautifully hidden among tall reeds a bittern covers its fluffy youngster while sternly surveying the photographer.*

northern hemisphere, with wingspans of almost 2 metres. The goliath heron of Africa is the largest species in the world. The small bittern of Eurasia hides deep in reed beds, where its streaked plumage provides excellent camouflage. The male bittern is famous for its booming call. The white cattle egret of Africa and Asia is so called because it is often found with grazing animals, picking up insects which are disturbed as the mammals feed.

The shoebill (*Balaeniceps rex*) is one of evolution's most bizarre creatures. Its huge beak is so heavy that the bird is obliged to rest it on its neck for most of the time. It is found only in the dense and extensive papyrus swamps of Africa, where it feeds mainly on lungfishes and catfishes.

The 17 species of stork are dignified birds that are almost voiceless. In courtship they communicate by clattering their bills and hopping up and down in rather ungainly dances. They are strong fliers and many migrate long distances. When flying, the neck is usually extended and the legs trail behind. Storks feed chiefly on fishes, insects and amphibians. The marabou and adjutant storks, however, also feed on carrion.

The Eurasian white stork (*Ciconia ciconia*) is regarded as a symbol of good luck in most countries and, in Europe, special platforms are sometimes erected on tall chimneys of houses to encourage these storks to nest there. They build great twig nests and return each year from their winter quarters in Africa.

The 28 species of ibis and spoonbill are medium to large in size. The ibises have down-curving bills, while the spoonbills have broad spatulate bills which are

Herons, storks and flamingoes

(ORDER CICONIIFORMES)

Most birds in this order have long, featherless legs with long toes for wading, a long bill, broad, rounded wings and a long neck. The large heron family contains over 60 species which are scattered over the world in temperate and tropical regions.

Most of them nest in colonies, building large stick nests in the tops of trees. Attractive courtship displays are often performed between partners, which look very similar. These birds feed in shallows, standing on one leg, or walking slowly in search of fishes or other aquatic animals.

The common heron (*Ardea cinerea*) of Eurasia, with its counterpart the great blue heron (*A. herodias*) of North America, are the largest species in the used for sifting food from the water in a side-to-side movement. The bright red plumage of the scarlet ibis of tropical America makes it one of the world's most striking birds. The plumage colour becomes more intense with age.

The flamingoes are very colourful birds that live in the tropics or subtropics of Europe, Asia, Africa and America. The strange shape of their bill is an adaptation for filtering their food from water. The bill is held upside-down in the water, and the tongue, working like a piston, forces water in and out over the specialized filtering structures. Plankton, crustaceans, worms and aquatic insects are obtained in this manner.

Flamingoes breed in colonies, often numbering thousands, usually on bare mud, in shallow water, or on islands in lakes. The nest is a low, flattened cone of mud, in which the female lays one chalky white egg. Both parents incubate the egg which hatches after about 28 days.

The greater flamingo (*Phoenicopterus ruber*) has the widest distribution, being found in India, Africa, the Mediterranean, South America and the Caribbean region. The lesser flamingo (*P. minor*) is found only in Africa; mainly on the soda lakes of the Great Rift Valley in East Africa. The Andean and the James' flamingoes are confined to the lakes of the South American Andes.

Below: *The Eurasian white stork is regarded as a symbol of good luck because it nests on houses and returns each year, after spending the winter in Africa, to the same large nest of twigs.*

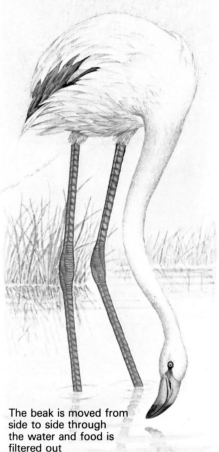

ADAPTATION IN BILL STRUCTURE OF FLAMINGOES

The beak is moved from side to side through the water and food is filtered out

Lesser flamingo

Greater flamingo

The lesser flamingo feeds nearer the surface than the greater flamingo and has a larger area of laminae

Laminae which filter out the minute organic food such as algae

Screamers, ducks, geese and swans

(ORDER ANSERIFORMES)

The 147 species of wild duck, goose and swan are found practically the world over. They are all water-birds with webbed feet, dense plumage, and most have flat bills with edges adapted for straining food. Most are strong fliers and many migrate great distances. The males are usually very colourful, often especially so during the breeding season, while the females are generally brown, a colour which provides good camouflage. Some of the species are illustrated.

Ducks include the familiar mallard, widgeon, shoveler, teal and pintail. These ducks are often called dabbling or dipping ducks because they dabble in shallow water, upended to filter-feed their food from the muddy water. Diving ducks include scoters, tufted and pochard ducks, all of which dive to the bottom of their aquatic home in search of the small creatures that live there and supply them with food. The shelducks are primitive ducks and are often so large that they

Above: *A mallard duck* (Anas platyrhynchos) *with her brood. They upend in order to filter-feed small animals from the mud with their beaks. Many of the ducklings will be eaten by predators.*

are misnamed geese, as is the case with the Egyptian goose and the Orinoco goose. The pochards or bay ducks include the canvasback, the redhead, the European pochard and several scaups. They do not have the bright metallic wing patches of the dabbling ducks. The perching ducks have long claws and a well developed hind toe and indeed they do spend more time in trees than any of the other groups of ducks. They include the colourful mandarin duck of eastern Asia and Japan

and the American wood duck. The widely domesticated muscovy duck is a perching duck. The eider duck is famous because of the habit of the female who lines her nest with the soft down from her breast.

The sea ducks are strong swimmers and divers, spending most of their life at sea but usually re

SPECIES OF DUCK

♂ – male
♀ – female

Mallard ♂

Shoveller duck

Mandarin duck

urning to fresh water to breed. They include the mergansers, which have a narrow bill with serrated edges, an adaptation for eating fishes. Sometimes these ducks are called saw bills. The goosander and the red-breasted merganser live around the North and South Poles. The stiff-tail ducks, which include the ruddy ducks of America, are small ducks with their legs placed so far back on the body that they waddle clumsily when on land.

The 6 species of swan are the largest of the waterfowl. They all have a long neck and feed on aquatic plants and small aquatic animals. Pairs often remain faithful for life, building a bulky nest in reeds or actually floating it on the water. Swans are white except for the black swan (*Cygnus atratus*) of Australia, and the black-neck swan (*C. melanocoryphus*) of southern South America. The white swans are the mute, whooper, trumpeter and whistling swans, all of which breed in the higher latitudes of the northern hemisphere. The mute swan is semi-domesticated and is the swan commonly seen on the lakes and rivers of city and countryside.

Geese are usually larger than ducks but smaller than swans. They spend much of their time on land, cropping and grazing grass rather as cattle do. When they migrate, geese fly in V-formation, keeping in contact with one another with the aid of their loud honking noises. The brent goose is the smallest species. It breeds in the arctic tundra and migrates south in winter. The Canada goose has been much reduced in number and range by extensive hunting. The greylag goose is the only goose to breed wild in the British Isles. The attractive snow goose of the northern regions of North America breeds on the edge of the Arctic Circle. The néné goose of Hawaii has been saved from extinction by breeding it in captivity, at, for example, the British Wildfowl Trust at Slimbridge in England. It is now being successfully reintroduced into its native islands.

The 3 species of South American screamer are fairly long-legged, wading, turkey-sized birds. They are not like ducks in appearance, but are definitely related to them.

Left: *The graceful black swan of Australia is often found in large numbers in big permanent swamps and lakes.*

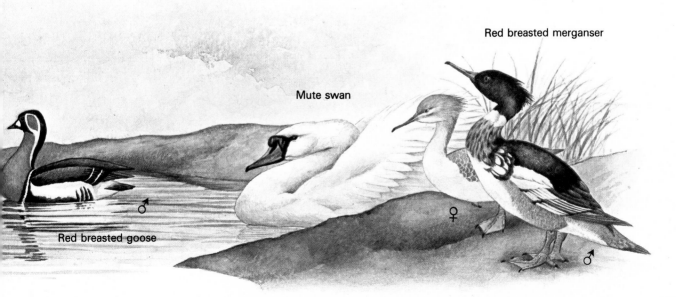

Red breasted merganser

Mute swan

Red breasted goose

Left: *An immature white-bellied sea eagle* (Haliaeetus species). *It will not reach maturity for several years. This impressive bird is closely related to the bald eagle, national emblem of the United States.*

Birds of prey

(ORDER FALCONIFORMES)

Birds of prey are recognized by their sharply hooked beaks with fleshy protuberances at the top. The group includes hawks, eagles and vultures, all of which have strong, powerful feet with an opposable hind toe. They feed in daylight, almost entirely on animal prey, although some are carrion feeders. Owls are often called birds of prey but they are classified separately. True birds of prey lay only a small number of eggs, usually 1 or 2, and thus have a slow reproduction rate.

The American vultures are the best soaring land birds, scanning the lands for carrion from great heights. They have such weak bills that their food must be part-

ly rotted before they can tear the flesh. The Andean condor (*Vultur gryphus*) and the Californian condor (*Gymnogyps californianus*) are among the largest living flying birds. Each has a wingspan of about 3 metres and weighs about 10 kilograms. The Californian condor unfortunately, is also one of today's rarest birds and it may actually be extinct. It has been shot just because it presented an irresistibly large target for hunters. The king vulture has a most strikingly colourful head. The black vulture and the turkey vulture are the best known American species and the most widely distributed of their family. They are often called buzzards.

The hawk family contains over 200 species, ranging from the quail-sized sharp-shinned hawk to the giant sloth-killing harpy

eagle of the Amazon jungles. Apart from hawks and eagles, the family includes kites, harriers and Old World vultures.

In such a large family it is not surprising to find a large variety in characteristics and habits. The members all have large, rounded wings, strong, medium-sized legs, and large hooked claws. They are usually brown or grey in colour, the plumage being barred or streaked. Normally the female is the larger of the sexes which otherwise look alike.

The goshawk (*Accipiter gentilis*), common in Eurasia but rare in North America, is typical of the small to medium-sized hawks. About 61 centimetres long, it has short wings, a long tail and longish legs. It breeds around the Pole in the northern temperate regions. Harriers are often called serpent eagles because they feed

168

Below: *An Andean condor soaring over its vast South American mountainous territory. It scavenges on any available food it sees, as well as killing fawns and lambs. It is said to dive out of the skies like a fighter plane.*

mainly on snakes and other reptiles. Unlike the majority of the birds of prey they often nest on the ground. Kites, as their name suggests, are soaring birds. The rare Everglades kite (*Rostrhamus sociabilis*) is also called the snail kite because it feeds only on one type of water snail, the apple snail (*Pomacea*).

The true eagles are the majestic members of the family, the best known being the regal golden eagle (*Aquila chrysaetos*) of Eurasia and America. It is now rare throughout most of its range because of shooting by man. It hunts rabbits, marmots, ground squirrels and woodchucks, in particular, grasping the prey in its huge talons. The nest or eyrie is usually high on a cliff or tree top and is enlarged annually by adding new twigs and sticks. The bald eagle of America is rare over most of North America with the exception of Alaska.

The vultures of the Old World look rather like New World vultures but anatomically they are related to hawks. The wingspan ranges from 1.3 to 2.4 metres, depending on the species. At dawn, once the warm air currents start rising from the ground, they are able to take off and then they soar on the thermals over their habitat for most of the daylight hours searching all the while for a kill or carcass. They live in the warmer parts of Europe, Africa and Asia. Each species is adapted to feed on a certain part of a dead animal, thus preventing competition between species in places where their ranges overlap. In East Africa, for example, the black, lappet-faced and griffon vultures have first sitting, while the Egyptian and hooded vultures wait their turn. The bald head and neck of some species is an adaptation which allows the head to be pushed inside a carcass without getting the feathers bloody and messy.

VULTURES – THEIR HIERARCHY OF FEEDING

White-backed vulture – feeds on the flesh of the carcass

Lappet-faced vulture – other vultures give way to this powerful bird which tears the carcass with its bill

The white-backed returns with a white-headed vulture, which has a naked head and neck to prevent body feathers becoming covered in blood.

Egyptian vultures pick the bones when the carcass is almost clean, and so have feathered heads and necks

169

Left: *A secretary bird of Africa feeds one of its young. Although it can fly, this strange bird prefers to run swiftly over the grasslands catching insects and even snakes, killing the latter by stamping on them with strong feet, or delivering killing strikes with its wings, which also shield the body from a biting snake.*

The fish-hawk or osprey (*Pandion haliaëtus*) has been given its own family. These birds are found almost world-wide, frequenting sea coasts or inland waters where they swoop down to fish. They hover above the water and, on spotting a fish, they dive, the large taloned feet being flung forward an instant before they enter water. Often they submerge completely to emerge clutching a trout or salmon in their talons. Their feet have pads with short, stiff, spicules that help hold the slippery fish.

The 62 species of the falcon family range from the small, darting falconets to the large gyrfalcon of the Arctic regions. Their

Left: *An osprey, also called the fish-hawk, feeds one of its 2 offspring. They catch fish in their talons from rivers or coastal waters. They fly above the water at a height of about 30 metres.*

flight is direct and swift and they do less soaring than other birds of prey. Most of them capture their prey with their feet after an aerial pursuit or in a fast air-to-ground dive. The kestrel, however, hovers 2 or 3 metres above the ground and then drops on an unsuspecting rodent or insect. The strangest member of the falcons is the caracara of South and Central America. It spends a lot of time on the ground on its long, strong legs. It feeds largely on carrion.

Falcons do not build nests. Some species, such as the gyrfalcon and the peregrine, lay their eggs on cliff ledges, while others such as the merlins, lay theirs in hollows on open ground. The number of eggs laid ranges from 2 to 5; the larger the bird the fewer the number. The eggs need about 4 weeks incubation, and this is usually carried out by the female. The young learn the skills of hunting from their attentive parents.

A most curious bird of prey is the secretary bird (*Sagittarius serpentarius*), which acquired its name from the crest which is reminiscent of a Dickensian bunch of quill pens stuck behind the ear. It runs after ground prey on its long legs, and often kills a snake by holding it in a foot and battering it to death with its wings, or by repeatedly dropping it from a height. The secretary bird is a strong flier, but prefers life on the ground.

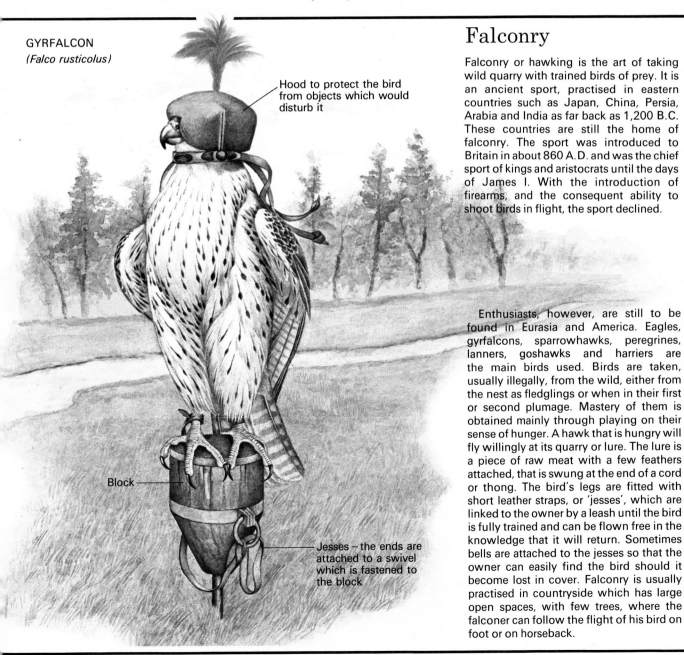

GYRFALCON
(*Falco rusticolus*)

Hood to protect the bird from objects which would disturb it

Block

Jesses — the ends are attached to a swivel which is fastened to the block

Falconry

Falconry or hawking is the art of taking wild quarry with trained birds of prey. It is an ancient sport, practised in eastern countries such as Japan, China, Persia, Arabia and India as far back as 1,200 B.C. These countries are still the home of falconry. The sport was introduced to Britain in about 860 A.D. and was the chief sport of kings and aristocrats until the days of James I. With the introduction of firearms, and the consequent ability to shoot birds in flight, the sport declined.

Enthusiasts, however, are still to be found in Eurasia and America. Eagles, gyrfalcons, sparrowhawks, peregrines, lanners, goshawks and harriers are the main birds used. Birds are taken, usually illegally, from the wild, either from the nest as fledglings or when in their first or second plumage. Mastery of them is obtained mainly through playing on their sense of hunger. A hawk that is hungry will fly willingly at its quarry or lure. The lure is a piece of raw meat with a few feathers attached, that is swung at the end of a cord or thong. The bird's legs are fitted with short leather straps, or 'jesses', which are linked to the owner by a leash until the bird is fully trained and can be flown free in the knowledge that it will return. Sometimes bells are attached to the jesses so that the owner can easily find the bird should it become lost in cover. Falconry is usually practised in countryside which has large open spaces, with few trees, where the falconer can follow the flight of his bird on foot or on horseback.

Fowl and pheasants

(ORDER GALLIFORMES)

The 240 or more birds in this large order all have the characteristics of the domestic fowl. Most are ground birds with short, stout, down-curved bills and large heavy feet which have 3 longish toes in front and 1 short one behind. These birds prefer to run and will fly only short distances. They cannot swim. They are worldwide in distribution, being absent only from Antarctica and Polynesia. The order includes the magapode, curassow, grouse, pheasant, guineafowl and turkey families.

The 165 species of pheasant vary from the 13 centimetre sparrow-sized painted quail to the giant argus pheasant and pea-fowl. The smaller quails and partridges are rather drab, but the pheasants include some of the most colourful and beautiful birds to be seen in the world. Several have been captured and reared in captivity for their beauty. Probably the most famous is the peafowl (*Pavo cristatus*) which originates in India and China. The peacock grows a superb display tail (actually long feathers from his back) which he raises to shimmer before the usu-

ally uninterested female. The pea-hen is dull brown coloured, as are most female pheasants, since this colouration provides excellent camouflage for a ground-nesting bird. The colourful jungle-fowl (*Gallus gallus*) of south-east Asia is the ancestor of the present day breeds of domestic fowl. The Ring-necked pheasant of Eurasia is now found across Eurasia and has been introduced into America. Most pheasants live in forest and scrub habitats. Males are polygamous and take no part in the rearing of their young. Courtship displays can be elaborate, with the males using their bright plumage to advantage.

Above: *Ancestor to the domestic fowl, a jungle cock and his hen study one another intently. They live in various habitats from sea level to 2000 metres and are members of the pheasant family.*

Grouse are of necessity well camouflaged, for the various species are preyed upon by larger birds of prey and by carnivorous mammals. Ptarmigans, for example, are a speckled brown in summer and so blend in with the rock and lichen habitat of their bare mountain or tundra home. In autumn, they moult and acquire a white plumage to blend with the winter snows. Grouse eat buds, seeds, and berries and during the breeding season they take insects

Silver pheasant
(*Gennaeus nycthemerus*)

Right: *The magnificent head of a male common turkey, better known as the gobbler. He spends much time displaying, shaking his wing feathers, raising his tail and grunting and gobbling.*

for their young. Courtship displays are often spectacular. For example, the capercaillie, the largest grouse, struts up and down emitting crackling and popping sounds, while the ruffed grouse drums on a log with its wings. The black grouse, sage grouse and prairie chicken all have communal display grounds. Here the males of the species perform as a group to attract the females.

Guineafowl originated in Africa. Many of the 7 species have tufts of bristly quills or bony growths on their bare heads. They were domesticated in Roman times. Today, they have been introduced into many countries where they are still hunted as game birds.

The wild turkeys from Mexico and much of the United States are fat and heavy birds that spend most of their time on the ground. They were domesticated by the Aztecs and introduced into Europe in about 1540. Turkey cocks are polygamous, courting females by raising their tail feathers, gobbling, and displaying their enlarged throat wattles and topknots.

The mallee fowl

The mallee fowl (*Leipoa ocellata*), is a large ground-dwelling relative of the pheasants that lives in Australia. It is a most interesting megapode (big-foot), providing a superb example of the way in which a bird has utilized natural heat-producing processes to incubate its eggs. Other megapodes are forest dwellers but this bird lives mainly in the open mallee and mulga areas of south and south-west Australia. It builds a huge mound, which can be over 1 metre high and almost 5 metres in diameter, from loose vegetation and loose earth. Under the earth the vegetation begins to rot and in doing so produces considerable heat. Over a period of about 6 days the hen lays from 15 to 24 large white eggs in a hole made in the top of the mound. The male adjusts his mound to keep the temperature to within one degree of 33°C during the long incubation period of 8 months. He uses his tongue and the sensitive lining of his mouth as a thermometer. On hatching, the chicks dig their way out of the mound unaided and are virtually neglected by their parents. They can fly within 7 days, but many fall to predators.

Sand layer to control heat loss from the mound

Egg chamber 33.3°C

MALLEE FOWL AND ITS NEST — DAYTIME

Sandy soil

Rotting vegetation to produce heat

dancing courtship displays in which 2 or more birds take part. The crowned crane (*Balearica pavonina*) of Africa, with its conspicuous topknot of stiff yellow feathers, has the habit of stamping on the ground so that its insect food takes to the air and can be spotted by the bird.

There are over 130 species of rail, many of them being solitary and secretive. All of them swim well and a large number spend most of their lives on and around water. There are several species that have lost the power of flight, probably because they live in areas where there are no enemies to disturb them. These include rails from New Zealand, Fiji, Hawaii, Jamaica and lonely Tristan da Cunha in the mid-south Atlantic Ocean. Well-known rails include the moorhen, the coot and the purple gallinule.

Cranes and rails

(ORDER GRUIFORMES)

The 14 species of crane are most attractive with their long necks, long legs and various adornments. They are usually brown, white or grey in colour, and the sexes are alike. Unfortunately the majority of species are now quite rare as a result of their being shot in large numbers during their annual migrations and also as a result of the destruction of their breeding and feeding habitats. The whooping crane, the Japanese or Manchurian crane, and the sandhill crane all survive in precarious numbers. Although all are now protected by law, they have become exceedingly rare.

All cranes are famous for their

Brown camouflaged plumage

Displaying

The great bustard

The great bustard (*Otis tarda*) has a bizarre and spectacular courtship display. Throughout the breeding season the male struts and stamps before the females, shaking his brown plumage vigorously and raising his tail over his back to expose the snowy white under-feathers. At the same time the wings are drooped and turned to show the white undersides of the flight feathers. This performance makes the bustard appear as a billowing mass of white. The display pose can be held for several minutes and then, with a sudden shake, the male transforms himself back once more into a brown camouflaged bird. As the females generally outnumber the males, a male usually mates with more than one female.

A flat or domed nest is built. It either floats on the water anchored to aquatic vegetation or is hidden in a reedbed. Both parents incubate some 2 to 16 eggs and care for the young.

The limpkin (*Aramus guarauna*) is a crane-like bird of America that behaves like a rail and is a link between the two families. It is interesting because its main food is snails and so the limpkin usually lives in swamps and marshes where water snails of the genus *Pomacea* are plentiful.

Bustards are also found in this group and are large, terrestrial birds that live on open grassy plains and brushy savannas of the Old World. Some 16 species are found in Africa. The kori or great bustard of the South African veldt is perhaps the heaviest of all flying birds, the male can weigh up to 23 kilograms. These bustards strut along on fairly long legs, feeding mainly on grain but also picking up insects. In Africa they are of great benefit because they feed on the locusts that plague agricultural lands.

Right: *A purple gallinule* (Porphyrula martinica) *demonstrating its lily-trotting habit, being supported by its very long toes.*

Waders, gulls and auks – the shorebirds

(ORDER CHARADRIIFORMES)

There are over 300 species in this large, diverse order. The 16 families fall roughly into 3 groups. The first group includes 12 families of shorebirds called waders. The second group combines the skuas, gulls and terns, while the third contains the auks, murres and puffins.

The plumage of nearly all the members of this order is dense and waterproof. Bright colours are rare, with black, white and various shades of greys and browns predominating. The sexes usually look alike, although the painted

Below: *A male ruff* (Philomachus pugnax) *fluffs up his collar on a lek, the communal patch of display ground where males display and attract mates.*

176

eft: A graceful shorebird, the avocet, its tiny organisms from the water with its slender and upcurved bill. The bill skims the surface of the water in a sideways sweeping movement.

nipes, phalaropes and the ruff show marked sexual differences.

Most species in the order breed only once a year, when 1 to 4 eggs are laid. The young, which are born after 2.5 to 4 weeks incubation, are downy and nidifugous (active soon after hatching and able to leave the nest almost at once). The parents usually share the responsibilities of incubation and rearing of the young, although with the snipe the female does most of the work, and with the phalarope it is the male who takes on these responsibilities.

The waders, gulls and auks and other members of the order are found practically all over the world, including the polar regions. The majority of these birds inhabit coastal waters, marshes, beaches and meadows.

Plovers, stilts and oystercatchers

The wader group includes the plovers, the long-legged stilts and avocets, the curlews, the sandpipers, redshanks, dunlins and phalaropes, the snipes and woodcocks, and the oystercatchers. All of these birds are migratory to a greater or lesser extent. Some only journey from their inland breeding sites to spend the winter on the coasts, while others migrate from their summer breeding sites in the arctic tundra to winter in tropical Africa. Many of the waders have characteristic cries, which are usually clear and wild, and often sound sad. For example the curlew's plaintive 'cur-lew' will carry far over the moors. The 'pee-wit' of the lapwing is another well-known call. The oystercatcher makes a shrill 'kleep-kleep' which changes to a 'pic-pic' when the bird is alarmed. Most of

the males produce a piping, trilling, nuptial song during their noisy courtship displays. Various displays are shown by the males of different species. The redshank and curlew raise their wings before the female, and then half lowering them they beat time with rapid quivers. They then take to the air hovering and uttering a bubbling song. The lapwing makes short erratic flights and is also very aggressive to other males. The male ruff grows a remarkable 'ruff' around his neck in spring, the colour of which may vary from bird to bird. The males congregate on a 'lek', a patch of open ground chosen as a courting site. Here the males engage in ritual dances and mock-combat displays. As the females, or reeves, outnumber the males, a male usually mates with more than one female but he takes no part in family affairs after this. Ruffs breed across northern Europe to Siberia.

Phalaropes

With phalaropes the roles of the sexes are reversed. The females are larger and more brightly coloured than are the males, and they take the initiative in courtship. The female grey phalarope (*Phalaropus fularius*) is called the red phalarope in summer because of her russet breast and body courtship plumage. After being

courted by a female and mating her, it is the male who goes off to build a nest, which is usually a neat cup in open tundra near to small freshwater pools. The female then comes and lays 4 heavily speckled olive-buff eggs, but it is the male who incubates them and tends to all the domestic chores. The females desert the nesting ground and gather offshore in small flocks, apparently unconcerned about their mates or offspring.

Female

Male

The curlews are the largest waders, varying in length from 53 to 66 centimetres. They inhabit marsh and moorland, where they use their long, slender, down-turned beaks to probe for insects and worms in soft soil or sand.

The sandpipers are smaller, more gracefully built birds. There are some 60 species to be found. The common sandpiper (*Actitis hypoleucos*) is a lively, restless bird, which lives on the banks of rivers or lakes, or on river estuaries. It continually bobs its head and jerks its tail. The redshanks and greenshanks are larger relatives that are easily identified by their long red or green legs.

The 56 species of plover have shorter, stouter bills than their wader relatives. They are swift-running ground birds, but they also have strong direct flight. Many species are active at night. The golden plover (*Pluvialis apri-*

caria) spends the summer in the Arctic tundra. After fattening themselves on crowberries in the North American summer range, the whole population flies non-stop to the northern coast of South America. From here they fly on to the wintering ground of the Argentine pampas. In January flocks begin to return northwards. This makes a round trip of more than 2,500 kilometres.

The ringed plover (*Charadrius hiaticula*), is one of several species that display the 'broken wing' trick. If an intruder advances towards her nest, eggs, or nestlings, the female will stagger away, trailing an apparently broken wing, thus offering the attacker an 'easy' prey. In this way the

Right: *Kittiwake family at home on their cliff nest site. The juveniles on the nest are called tarrocks. The black band at the nape will disappear by the first summer and adult plumage is shown by 2 years. Tarrocks often gather in autumn flocks.*

Left: *A lapwing of Eurasia settles down to brood its 4 spotted eggs which hatch in about 27 days.*

danger is drawn away from the crouching chicks, or the eggs. When the female has lured the intruder sufficiently far away she takes to the wing, leaving a perplexed, still hungry enemy.

The stilts and avocets are distinguished from the other waders, by their very long legs, hence the common name of stilts. The long legs enable these birds to wade in deeper shallow water than other waders, and here they use their long bills to feed on snails, crustaceans, insects and sometimes fishes and vegetable matter.

Gulls, skuas and terns

There are about 90 species of gull, skua and tern. Best known of all seabirds, gulls are strong fliers,

often soaring and gliding. Some gulls will often venture a long way inland and breed on inland waters. The herring gull (*Larus argentatus*), for example, is often seen in large flocks on rubbish tips or following tractors in search of food. Most of this group are white in colour although the skuas are brown. They all have long wings and webbed feet. They generally nest in colonies, some species choosing ledges of cliffs, while others prefer sand dunes, coarse grass or even swamps.

Above: *A common tern* (Sterna hirundo) *presents a fish to his mate, an important part of their courtship. The fish is obtained by shallow diving or by snatching while the tern is in flight.*

The smallest gull is the little gull at 28 centimetres, while the largest is the glaucous or greater white-winged gull of the Arctic, which is 81 centimetres long. The herring gull which seems quite large, is a mere 58 centimetres long. This bird is circumpolar in the northern hemisphere and is the commonest gull on the Atlantic coasts of North America and Europe. Gulls are quite aggressive birds and have been known to invade tern nesting colonies and to drive the terns completely away within a few years. However, when the gulls move off to new breeding grounds, gradually the terns come back. Gulls usually lay 2 to 3 eggs and both parents share the family duties. Terns are smaller and slimmer than gulls, and have relatively longer tails and wings. Most species will dive into the sea for fishes, a habit not shown by most gulls. They are often called 'sea swallows' because they are so graceful. Their nest is practically non-existent, the 2 or 3 eggs being laid on a shingle bed or in sand dunes or dry grass. They defend their 'nest' with boldness against other members of their breeding colony. The Arctic tern is famous for its tremendous migratory flights. The dusky black noddy is a tropical species.

The 4 species of skuas are the pirates of the polar regions. They have strong, hooked claws and sharp, hooked bills which they use to attack other birds and make them disgorge their food, or to kill the nestlings of other birds, which are usually their relatives – the gulls and terns. Skuas are famous for their attacks on visiting scientists intruding into their breeding grounds.

Close relatives of the terns are the 3 species of skimmer. They have remarkable beaks, the upper half being shorter and more flexible than the lower half. This 'scissor bill' is used to catch food as the bird skims over the water, the lower half of it just scooping the surface.

Below: *The Arctic skua* (Stercorarius parasiticus) *is a pirate of the northern polar regions. It attacks other birds for their food, or kills young nestlings, and also scavenges along the shore for carrion.*

Sombre auks and colourful puffins

Auks, guillemots, razorbills and puffins all live in the cold regions of the northern hemisphere. They all have heavy, compact bodies with the legs placed far back, and they have webbed feet. They feed on small fishes and crustaceans, and breed in colonies on rocky coasts, spending much of their time out at sea. They usually lay a single egg.

The largest member of this family was the flightless goose-sized great auk, which became extinct in the nineteenth century, mainly because it was persecuted for its feathers. The largest living auks are the common auk, Brunnich's auk or the thick-billed guillemot or murre, and the razorbill. These birds breed on ledges on seaward facing cliffs. The large egg is pear-shaped, an adaptation that prevents it rolling off the narrow ledge.

The puffins, with their gaudy parrot-like bills, are the clowns of the family. The 2 Pacific species, the horned and the tufted puffin have additional head decorations.

The common puffin (*Fratercula arctica*) develops its multi-coloured bill during the breeding season. The sexes are alike and the bill is most important in keeping the bonded pair together. After the breeding season, the puffins moult their feathers and the outer covering of the bill is also shed. They are quite dull in appearance in the winter months with their black and white plumage and ordinary yellow bills.

Opposite page. Top, left: *A puffin demonstrates its remarkable skill of holding several fishes in its gaudy beak. Up to 11 fish can be held at one time.* Top, right: *A razorbill* (Alca torda) *can also hold fishes in its beak. They dive into the sea to obtain them and swim underwater using their wings.*

Young birds

There are two types of young bird. At hatching the baby bird may emerge blind, with little or no down, and be totally dependent on its parents for food. These are known as altricial birds but they still have the instincts necessary to ensure their growth and survival. Thus they open their mouths wide, showing bright colour inside – the gape – which stimulates the parents to provide them with an almost constant supply of food. During the first few weeks of their life they live in the warmth and shelter of the nest. This has given these chicks the alternative term nidicolous, which means nest-attached. Nestlings of all perching birds, such as blackbirds, tits and thrushes, as well as woodpeckers, kingfishers, hummingbirds, parrots and pelicans, are altricial.

The other kind of baby bird is one which is alert, covered with warm down, and able to run about at once. These chicks are called 'precocial' and are usually very attractive. Just think of the domestic day-old-chick and this gives a good idea of what these babies are like. Precocial chicks are found in most ground-nesting birds where it is important for the young to be able to move about immediately in order to escape from hunting predators, such as foxes, snakes and birds of prey. Examples include the chicks of pheasants, quails and other game birds, those of rails, and those of waders such as plovers, curlews and lapwings. These chicks leave the nest as soon as the last of the brood has dried, and thus they are nidifugous, which means nest-leaving, although many of them stay with the parents and near the nest site for several days or weeks, until they can fly.

Blackbird nestling
(altricial) – 3 days old

Blackbird a few
days older

Lapwing chick
(precocial) – 3 days old

Pigeons and doves

(ORDER COLUMBIFORMES)

About 290 species of pigeon and dove are distributed world-wide, with the exception of extreme latitudes and some oceanic islands. The smaller species are often called doves and the larger ones pigeons. The soft, dense plumage is often highly coloured and patterned, especially in tropical pigeons and doves. They feed mainly on seeds, buds and fruit, and are often shot in great numbers because they plunder agricultural crops.

Tropical species include the brilliantly coloured orange dove of the Fiji Islands, the magnificent fruit pigeon of New Guinea and northern Australia, and the crested and bronze-winged doves of Indonesia and Australia. The largest pigeons are found in New Guinea, these being the crowned pigeons, splendid birds with erect, fan-shaped crests. The tooth-billed pigeon of Samoa is remarkably like the extinct dodo of Mauritius. The dodo was a flightless, swan-sized bird that became extinct by the end of the seventeenth century, due to thousands being killed for food by sailors. The passenger pigeons of Central and North America are now also extinct, having been slaughtered in their millions by hunters. The gentle cooing woodpigeon of Europe is an enemy of most farmers, since it destroys his crops.

The rock dove is the species from which most of the domestic breeds such as fantails, tumblers and pouters have been bred. In the truly wild state they nest chiefly on rocky cliff faces.

Right: *The dodo is probably the most famous extinct bird. This flightless pigeon now exists only as a few museum specimens and skeletal remains.*

The nest of most species is usually an untidy, flimsy structure of twigs sited in a tree or on a ledge, although a few nest on the ground or in a burrow. Usually 1 or 2 eggs are laid and the parents share the incubation duties. The naked, nidicolous young are fed for the first few days on pigeon milk, a rich yellow liquid produced in the parent's crop. Later on they receive regurgitated solids.

Right: *Feral pigeons in Trafalgar Square, London. Their cooing voices are familiar sounds in cities and towns, where they survive on bread, cake and bird seed offered by the public, and scraps raided from rubbish and litter bins.*

Twentieth century invasion

is unusual in these days when man ersecutes so many wild animals to the oint of their extinction, to find certain pecies increasing their range and num- ers. One such case concerns the collared ove (*Streptopelia decaocto*). Early in the wentieth century it had slowly spread om its homeland in Asia to Bulgaria. ince then it has crossed Europe to the est at an explosive rate. In the 1930s the ird had reached Yugoslavia, and by 1947 airs were seen in the Netherlands. A year ter the species had reached Denmark and n the following year, pairs were nesting in weden. This bird had reached Belgium by 952 and was first sighted in Britain in the ame year. The first nesting pair was corded in Norfolk in 1955, and now the pecies is widespread in Britain. It may ven reach America because exhausted irds have been found more than 80 ilometres west of Britain. With favourable vinds they could perhaps reach and estab- sh populations in America.

One reason why this 28 centimetres ng dove has done so well in invading the west is that it is extremely adaptable. It will nest in bushes and walls, as well as in its preferred pine forests. Up to 5 broods a year help to boost the population so that the dove becomes more widespread.

Above: *The collared dove, a recent arrival in Europe and Britain. Since the beginning of this century it has steadily advanced westwards from the Balkans across Europe.*

Strange parrots of New Zealand

New Zealand is the home of 3 strange parrots. These are the kea (*Nestor notabilis*) and the kaka (*N. meridionalis*) which are both fairly large with greenish plumage, and the smaller, rare kakapo (*Strigops habroptilus*).

The kea has been branded a sheep-killer as a result of isolated and circumstantial evidence. Consequently it was persecuted on its mountainside home in the South Island. It certainly feeds on carrion and will perhaps attack an ailing or injured mammal, but it is scarcely credible that this parrot could kill a healthy mature sheep. The related kaka is distinguished by its scarlet abdomen and its preference for a forest habitat rather than open ground.

The kakapo is now very rare because it has had to compete with introduced deer for food and living space, and to contend with introduced predators such as dogs, cats, weasels and rats. Furthermore much damage has been done to its habitat of virgin forest. It seems that this bird was naturally declining several hundred years ago but its disappearance has been accelerated by man and his impact on the New Zealand islands. This curious ground-dwelling parrot does not fly but scampers about on its beech forested hillsides with great agility. It can also hop and glide. Today the only known population of this nocturnal vegetarian is at the Cleddau watershed in Fjordland on the South Island.

Below: *A kea.*

Parrots and parrakeets

(ORDER PSITTACIFORMES)

The 500 or more species in t order are familiar, instan recognizable birds with sho stout, strongly hooked beaks a clinging feet where 2 toes po forwards and 2 backwards. Th are long living birds and are of brilliantly coloured. They inha mostly tropical and subtropi regions. The order includes p rots, macaws, lovebirds, coc toos and budgerigars.

Parrots vary greatly in size, great black cockatoo of N Guinea and northern Austra being the largest, at about centimetres. Its bare fac patches change from pink to with the bird's chang emotions. It prefers to live alc or as one of a pair, unlike Australian relatives, the sulph crested and rosy cockatoos, tl fly together in huge flocks. T small, well-loved budgeri; (*Melopsittacus undulat* originates in Australia wh wild birds are green. The bl

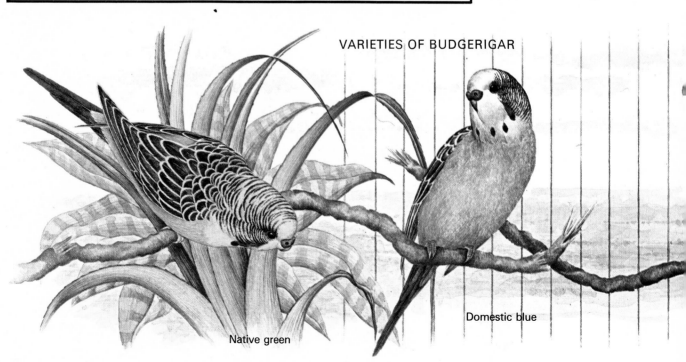

VARIETIES OF BUDGERIGAR

Domestic blue

Native green

yellow and white varieties have been bred in captivity.

The best-known parrot is the African grey parrot with its grey plumage enhanced by a striking red tail. The vivid little lovebirds also live in Africa and Madagascar. Their 'kissing' behaviour in captivity is always amusing to watch. Tropical America boasts the most species, and these include the large, long-tailed macaws. The brilliant scarlet macaw and the gold-and-blue macaw are almost a metre long, but two thirds of this length is tail. These birds have bare faces and massive hooked bills with which they can crack nuts, such as Brazil nuts, without difficulty. All macaws (*Ara* species) are noisy and showy birds. The Amazon parrots inhabit the same range as macaws. They are mainly green with bright markings of red, yellow or blue.

Right, above: *A sulphur-crested cockatoo* (Cacatua galerita) *on the alert. These noisy birds of Australia fly in huge flocks, often raiding crops of peanuts and maize.*
Right, below: *The hyacinth macaw* (Anodorhynchus hycacinthus) *lives only in the tropical jungles of interior Brazil. It has a massive, strong, hooked beak.*

Domestic clear body opaline

Cuckoos and touracos

(ORDER CUCULIFORMES)

When the cuckoo is mentioned the first thought that usually comes to mind is either of its famous call that heralds Spring in Europe, or of the fact that the female bird lays her eggs in other birds' nests and relinquishes all parental care. These facts are true but not for all the 128 species of cuckoo which are found all over the world. The parasitic European cuckoo is probably the best known. It is rather drab with its bluish-grey plumage, and measures about 30 centimetres in length. The 'cuck-oo' is usually the only indication of the bird's presence, it is seldom seen.

The numerous tropical species of the order, however, are much larger and often quite gaudy. The touracos, or plantain eaters, of the forests of tropical Africa, for example, are almost a metre long and have brilliant green and blue plumage with crests adorning the head. A remarkable feature of the 18 species is the bright red wing plumage, seen in flight and display.

The really fascinating aspect of cuckoos is the parasitic habit of the majority of species. The female European cuckoo (*Cuculus canorus*), having mated, selects the nest of another bird, such as the meadow pipit, reed warbler, sedge warbler, hedge sparrow or wagtail. Watching carefully, she times her manoeuvre so that she is able to quickly lay 1 egg while the nest owners are away briefly and before the chosen host female has started to incubate her own eggs. The female cuckoo may lay

Left: *A foster parent reed warbler, looks at its enormous foster nestling, a cuckoo. The chick gets all the attention having disposed of other nestlings or eggs on hatching.*

Cuckoo chick removing eggs from a reed warbler's nest

12 to 24 eggs in different hosts' nests, but she usually chooses nests belonging to the same species of foster parent. The foster parent is usually unaware that there has been an intruder. Quite often the cuckoo removes an egg of the host and transports it away or swallows it.

The young cuckoo hatches out at about the same time as the young of its foster parents, and within hours it instinctively gets rid of its fellow nestlings or the unhatched eggs. Although blind, naked and otherwise helpless, it is born with the knowledge necessary for survival. It will laboriously get each of the other eggs or nestlings, in turn, on to its back and then jerk them over the brim of the nest. In this way the baby cuckoo secures the whole attention of its foster parents, who still do not recognize it as an intruder. They expend all their energy in finding enough food to feed the extremely hungry youngster. It grows rapidly and soon fills the nest, quite often being so large that the foster parent stands on its back or head to pop food into its gaping mouth. It is able to fly in about 4 weeks but is fed for a further 2 weeks. Then it abandons its hard working foster parents. An incredible feat is then achieved by the young cuckoo, which is able to migrate to tropical parts of Africa, following a route that the adult birds have taken much earlier. It obviously finds its way purely by instinct.

The American cuckoo is not parasitic on other birds. It builds its own nest, crow-like, high in a tree, where the eggs are incubated by both parents.

Roadrunners of the American south-western deserts are ground cuckoos that have long, sturdy legs. They run and catch small snakes and lizards, which they kill by pounding them with their heavy bills. They then swallow the prey head first. They are weak fliers, preferring to run rather than take to the air. They have been recorded travelling at over 35 kilometres per hour, on the ground.

Below: *A roadrunner of the southwestern deserts of America. This ground dwelling cuckoo can fly but prefers to run rapidly along the ground, searching for small lizards, insects or even snakes. It kills the latter by pounding them with its bill.*

Owls – nocturnal birds of prey

(ORDER STRIGIFORMES)

Owls are nocturnal birds of prey, as opposed to the eagles, hawks and their relatives that hunt during daylight hours, and are thus diurnal birds of prey. Owls are soft-plumed, short-tailed, big-headed birds with large, direct eyes directed forwards and surrounded by a facial disc. The bill is hooked and directed downwards and does not look very strong when compared with an eagle's beak, for example. The beak is usually partly hidden by feathers of the facial disc. The claws are sharp and strongly hooked for dealing with prey.

In size, owls range from a little larger than a sparrow, for example the elf owl and the pygmy owl, to almost that of an eagle, for example the eagle owl.

The barn owl (*Tyto alba*) is distributed worldwide with the exception of the polar regions. There are several subspecies, but in general the bird is a nondescript buff and brown with an attractive white face. It often hunts on nights too dark even for its keen night vision. Then it depends heavily on hearing to locate the small ground animals, such as shrews, mice, voles, rats, insects and small birds, on which it feeds. It swoops swiftly down on its victim, often giving a prolonged strangled shriek.

The barn owl does not build a nest, but lays the 4 to 6 eggs on a heap of pellets made of the indigestible fur, feathers and bones of its prey. The female incubates the eggs for about 33 days, but both parents feed the young. These can fly after 9 to 12 weeks.

In many parts of its range the barn owl has decreased in numbers, due mainly to the effects of agricultural programmes, and the

loss of nesting sites, such as abandoned buildings and hollow trees.

The other 123 species of owl display various adaptations according to where they live. The little owl (*Athene noctua*) of Eurasia usually nests in tree holes but will happily take over an abandoned rabbit hole. The burrowing owl (*Speotyto cuncularia*) of the prairies of North America nest in ground holes, usually taking possession of abandoned burrows of prairie dogs, badgers, skunks and ground squirrels. However, where necessary it will enlarge the hole by scraping away the earth with its powerful feet and kicking it into a mound outside. It will excavate its own hole in sandy areas such as in Florida.

Long eared owl clutching prey

Left: *A silent ghost of the night, a barn owl leaves its hole on a hunting trip. It can see very well in the dark but also relies on excellent hearing, being able to find prey such as a mouse by sound alone.*

189

Left: *A female snowy owl crouches protecting her young chicks. Although they feed mainly on lemmings, they are also quite skilled at catching fish, seizing them in their claws.*

The beautiful snowy owl (*Nyctea scandiaca*) lives in the Far North were there are no trees. It places its nest on a rise on the tundra, so that it can keep a watch over the landscape. Its breeding success is closely linked to that of its prey, which consists largely of lemmings. When lemmings undergo a population explosion, which happens usually every 4 years, the owls produce more youngsters, since there is an abundance of food. When the lemming population crashes the owls are faced with a serious food shortage and may have to move southwards to temperate regions, and the following nesting season a female will only produce 3 or 4 eggs. So the number of eggs ranges from 3 in lean years to as many as 12 in good years. Female owls, like female hawks, are usually much larger than their mates, and as a rule, it is only the female that incubates the eggs. All owls are very courageous when defending their nest. Even small species will attack a man if he ventures too close to a nest. The female begins to incubate with the laying of the first egg, and as a result the babies hatch out at intervals of 1 or 2 days. This is known as 'staggered birth' and is thought to be linked with food availability. If food is plentiful then all the chicks will be fed, but if food is scarce the first to hatch will be the strongest and they will receive the most food and so will survive rather than the later hatching nestlings. Baby owls are most attractive, being covered in fluffy, dense white down.

How to tell what an owl has eaten

An owl may eat 3 or 4 animals, such as small rodents and birds, in a night. All of the animal is swallowed so that inside the bird's gizzard there are large amounts of indigestible bones, fur and feathers. This is compacted into a tight pellet and regurgitated by the owl. The bird usually eats at its roost, and so, if a search of the ground reveals owl pellets, usually there will be an owl's roost in the tree or building above.

By dissecting a pellet it is possible to find out what the owl has eaten. Parts of shells, ribs, teeth, backbones, jaws, limb bones, girdles and so on, can all be separated using fine tweezers, needles, and a lot of care and patience. If the pellet is very dry it may be necessary to soak the broken pellet in a jar of water for several hours. When all the small bones have been separated they can be whitened by bleaching them in hydrogen peroxide for a few minutes. Great care should be taken with bleach and none should be spilt. The bones should be washed in clean water afterwards. Then after drying them and arranging them in the various groups, they can be stuck on to a card.

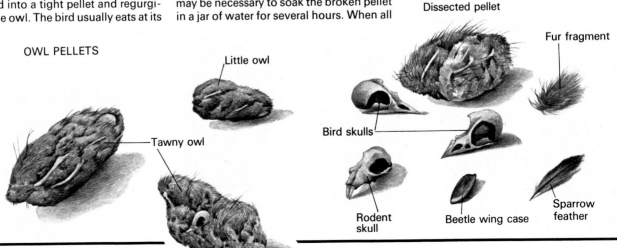

OWL PELLETS

Little owl

Tawny owl

Dissected pellet

Fur fragment

Bird skulls

Rodent skull

Beetle wing case

Sparrow feather

Hunters of the city

Four birds of prey have become adapted, more or less successfully, to city life. These are the kestrel, tawny owl, great horned owl and the black kite. Of these, the tawny owl (*Strix aluco*) is remarkable in that it has had to change its diet in order to survive successfully in the city. In the country, over 90 per cent of its diet is made up of small mammals, the remainder usually consisting of small birds. In towns, the long grass where small rodents abound is lacking, and so the proportions are reversed. Some 90 per cent of the city owl's food is house sparrows and starlings, the remaining 10 per cent being made up of small mammals. In towns and cities the tawny owl has adapted to nesting in the hollows of trees in parks and gardens, rather than in its normal woodland home.

The great horned owl of North America (*Bubo virginianus*), unlike the related eagle owl of Eurasia, is tolerant of man and occasionally nests in city parks. The deep hooting of this bird can be heard within a few kilometres of the heart of New York City. In towns the kestrel (*Falco tinnunculus*) has successfully bred on the window ledges of high-rise flats or office blocks. It feeds mainly on small rodents, house sparrows and starlings, which it catches by hovering above them and then swooping quickly down upon them.

The black kite (*Milvus migrans*) of southern Asia, is an efficient municipal scavenger, regularly visiting garbage dumps, slaughter-houses and meat and fish markets, where there are easy pickings to be had. It will also poach from poultry farms. In flight it is a striking bird with its forked tail. It is often seen in seaports, towns and cities throughout southern Asia, where it will perch on telegraph poles and roof tops, scanning its surroundings for titbits. It uses its expert flying techniques to dodge the overhead cables and the various vehicles of the busy oriental roads and streets.

Above, right: A kestrel hovers over a street lamp-post on the look-out for small rodents, or house sparrows. It nests quite readily on window ledges of buildings.
Right: A black kite perches on an industrial pipe. This large bird of prey is widespread over south-east Asia and parts of Australia. It eats any kind of edible waste. Flocks of hundreds gather at garbage dumps, slaughter-houses or markets. They will follow tractors in a similar fashion to northern hemisphere gulls, swooping down to take disturbed insects.

Nightjars, goatsuckers and potoos

(ORDER CAPRIMULGIFORMES)

The 90 or so species of this order are all nocturnal birds with long pointed wings, small feet and small bills that open into tremendous gaping mouths. The name goatsucker originates from an ancient fallacy that at night these birds milked goats with their huge gaping mouths. This huge gaping mouth is in fact used by all birds of the order, except the oilbird, to trap insects while on the wing. The birds hide by day and so are sombrely coloured in browns and greys, with various patterns to help them merge with the leaf litter of the forest floor or against background trees.

The strangest goatsucker is the oilbird (*Steatornis caripensis*) of northern South America and Trinidad. Because of its piercing shrieks and calls, the Spaniards call it *guachara*, which means the

Above: *Resting oilbirds are caught in the photographer's flash. These birds have a hooked beak like a hawk and also certain owl-like characteristics.*

one that wails or cries. It is the only vegetarian in the order, leaving its pitch dark caves at dusk to

Male nightjar feeding on insects

Right: *Perfectly camouflaged by its mottled plumage, a gaboon nightjar lies hidden on the forest floor. Its nest is a mere scrape in the vegetation and soil.*

feed on the oily fruits of palm trees. These may be over 80 kilometres away, but just before dawn they streak back to their caves. Another astounding fact is that these birds can use echolocation as bats do, to find their way about.

The potoos are large goatsuckers found in South America and the West Indies. They perch upright on a broken shrub or branch, sitting perfectly still so that from a distance they are not noticeable, since they look like a continuation of the tree. The frogmouths are Australian members of the family, and they differ from the other goatsuckers in their method of feeding. They prey on creeping, crawling things, such as ground beetles, centipedes, scorpions and caterpillars. The big tawny frogmouth of Australia will also catch marsupial mice.

There are some 67 species of nightjar scattered throughout the temperate and tropical parts of the world. Their name is a result of their loud, persistent monotonous calls, which 'jar' the night. Individual nightjar specific names indicate the sound of their calls. For example there are whip-poor-wills and chuck-will's-widow.

Nightjars do not build a nest, but the female lays 2 eggs on the bare ground. She sits on them and relies on her dappled brown colouration for protection. The male helps with the task of incubation and the rearing of the young. In North America the common nighthawk has taken to nesting on the flat roofs of town buildings.

Right: *A chuck-will's-widow sleeping. This North American nightjar gets its name from its call. It frequents wooded marshes and rocky hills in the western United States and winters in the West Indies.*

Dazzling flying jewels

The majority of the 319 species of hummingbird found mainly in tropical America are so brilliant and gem-like that they were used in the nineteenth century for jewellery. Millions were caught, skinned and transported to European and North American cities, where they were fashioned into pins, brooches and other fashion objects. Thankfully the fashion changed and most species of hummingbird are no longer in danger of extinction.

With their small size and expert flying skills, hummingbirds are well adapted to invade the world of flowers. The sweet, sugary nectar from flowers gives them tremendous energy. They also eat the tiny insects taken from the flower petals. To obtain the nectar they have long, thin beaks and long tongues. Some have a tongue that is tubular for sucking out the nectar, while others have a tongue with a brushy tip for collecting the nectar and obtaining the pollen and insects.

Hummingbirds can dart, stop and start, hover, move up, down, sideways and even backwards. All this time their wings never stop beating, seeming just a blur to the human eye. The sound produced as the wings beat is responsible for the name hummingbird. In addition to the vivid array of colour—iridescent greens, reds, yellows, purples and blues — hummingbirds have various crests, fans, and tail adornments for use in courtship. The male is very territorial, guarding a space of a few square metres against other male hummingbirds or even other avian intruders. He performs a dashing aerial courtship display for a female and they often mate in flight. The female is then ignored by the male. She builds a tiny, deep, cup-shaped nest of plant down and spider's webs, usually in the fork of a branch. Here she lays 2 pure white eggs and feeds the young when they hatch by pumping food from her stomach into their throats.

The smallest bird living today is a hummingbird. This is the tiny 5.4 centimetre bee hummingbird. Its minute nest is a mere 2 centimetres in diameter. The largest hummingbird is the 21 centimetre giant hummingbird of the high Andes, whose wingbeats are slow enough to be seen by the human eye. The ruby-throated hummingbird ranges up eastern North America, flying 800 kilometres across the Gulf of Mexico without stopping. How it manages to survive without 'refuelling' mystifies scientists.

Above: *A hummingbird caught in action. The body is held quite still in the air while the wings vibrate at up to 100 beats per second. A tubular tongue is extended from the long beak to suck up the sweet nectar from the flower head.*

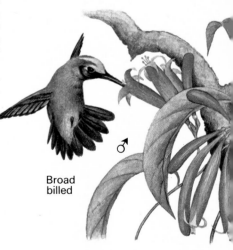

Broad billed

♂ — male
♀ — female

Ruby throated ♂

ne mmingbird

Adorable coquette ♂

Ruby roated estling

Ruby throated ♀

Broad tailed ♂

Swifts and hummingbirds

(ORDER APODIFORMES)

Swifts and hummingbirds are without question the kings of flight. Their wings have developed at the expense of their feet. The order's scientific name Apodiformes means 'the footless ones' and, although they are not footless, their feet and limbs are tiny.

The aptly named swifts are very fast fliers. The spin-tailed swifts of eastern Asia are the fastest, being able to fly at over 240 kilometres per hour. Swifts live on insects that they catch in flight and can often be seen diving and swirling in the swarms of flies and gnats, on warm summer evenings. Swifts use their tiny clawed feet to cling to upright surfaces, such as cave walls, hollow trees, towers and chimneys, where they nest and sleep. On the ground they become stranded; the legs are so short and the wings so long that they are unable to take off. They bathe and drink by swooping in and out of the water. Scientists have found that some swifts, including the European species, spend the night on the wing, high in the air, not once returning to the nest.

Swifts secrete a glue-like saliva which is used to cement straw, feathers or grasses together to form a nest. The cave swiftlets of southern Asia make their cup-shaped nest out of their gluey saliva, adding little or no other nesting material. Natives use long bamboo ladders and poles to reach the nests and the Chinese use them to make bird's nest soup, a delicacy which is very expensive.

Below: *A European swift catching flies. The wings are so long and the legs so short the bird is stuck if grounded.*

Kingfishers and hornbills

(ORDER CORACIIFORMES)

Kingfishers, of which there are 87 species, are found the world over. With their long, slender bills, they are usually fish-eaters. The European kingfisher (*Alcedo atthis*), with its brilliant blue plumage, is typical of the kingfisher family. It sits on a chosen perch watching the water beneath. Then it flashes into the stream, swimming underwater with its wings, and usually emerges with a small fish, such as a minnow or trout. Returning to its perch, it tosses the fish and catches it in

such a position that it can be swallowed head first. This prevents the spines and scales from sticking in its throat.

A pair of kingfishers excavate a long burrow in a soft soil bank and both parents rear their 2 to 7 youngsters. The nestlings' burrow is incredibly filthy, strewn with fish bones and the remains of food. The young's feathers are protected by waxy sheaths, which are shed just before they are ready to leave the burrow, and fly.

The laughing jackass kingfisher or kookaburra of Australia is famous for its laughs and chuckles. It is called the bushman's clock because of its dawn and evening choruses. It feeds

Above: *A kingfisher leaves its nesting hole excavated in a soft sandy bank above a river. The tunnel is up to a metre long and thinly lined with discarded fishbones. Six or 7 white eggs are laid.*

mainly on lizards, snakes, large insects, and small marsupial rats and mice.

The 45 species of hornbill live in the tropical forests of Africa, Asia and some South Pacific Islands. They are identified by the massive downward curved bill and many have a horny casque growing up from the beak and head. Most of the hornbills live in trees and feed on fruit and insects.

During the breeding season the female is imprisoned in a tree hole by her mate who blocks up the

entrance with clay or dung. She lays between 1 and 6 eggs which she incubates for 30 to 50 days. She and the young are fed by the male, the food being passed through a small opening. Usually the female is released before the young are fledged. The pair then replaster the hole and tend the young birds together until it is time for their freedom. The hoopoe, rollers, bee-eaters and mot-mots are related to hornbills and kingfishers.

Above: *An Indian hornbill displays its massive bill and horny casque. In its jungle home it eats ripe fruit and insects it finds among the foliage.*
Right: *A laughing kookaburra (Dacelo gigas) holds a lizard in its stout beak. This bird is a kingfisher that no longer fishes.*

Woodpeckers and toucans

(ORDER PICIFORMES)

Woodpeckers and their relatives the toucans, barbets, honey-guides, jacamars and puffbirds all have the same kind of feet. Two toes point forwards and two face backwards, this type of foot being named zygodactylous (yoke-toed). They share this feature with parrots, cuckoos and trogons.

There are over 200 species of woodpecker, which are found all over the world except for the polar regions, Australia, Madagascar and a few oceanic islands. They are tree-dwellers and are well known for their habit of boring holes into trees to make a nest hole or in order to obtain insects and grubs. The bird clings to a tree trunk, using its stiff tail as a prop and its powerful, long, straight bill as a hatchet, chisel, or borer. The tongue of a woodpecker is long and worm-like and can be extended far beyond the bill. Its barbed tip harpoons grubs or insects so that they can be pulled out of the tiniest tunnel. A nest hole is always smooth and well finished inside, and in it the 2 to 8 eggs are laid on the bare wood.

The 37 species of toucan look rather like hornbills and are found only in the tropical regions of the New World. The hugely enlarged bill is often more brightly coloured than is that of a hornbill, as for example in the toco toucan and the keel-billed toucan. Despite its size, the bill is light and capable of delicate manipulations. The long fringed tongue is used in conjunction

Left: *A great spotted woodpecker* (Dendrocopus major) *at its nesting hole. Two toes of each foot face forwards while 2 dig in backwards, typical of all members of its family. The tail is used as a prop when it clings vertically.*

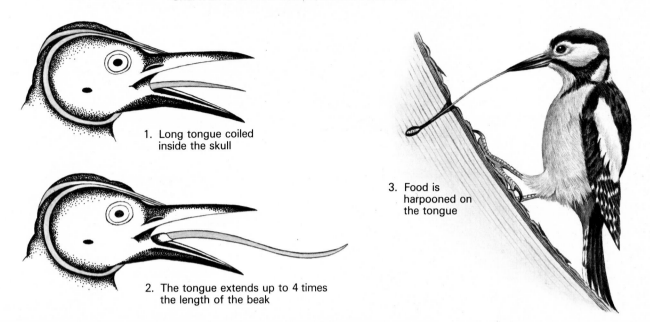

1. Long tongue coiled inside the skull

2. The tongue extends up to 4 times the length of the beak

3. Food is harpooned on the tongue

with the bill to enable the bird to feed on berries, fruit, insects and spiders. The larger species such as the toco and Cuvier's toucan are known to take eggs and nestlings.

The aracaris are small toucans that live in small flocks throughout the Amazonian jungle. They are quite colourful, as also are the hill toucans (*Andigena* species), which move up and down the Andean mountainsides as various fruits and berries ripen.

Toucans and aracaris nest in tree holes, often enlarging a hole previously pecked out by a woodpecker relative. Most species leave the hollow bare and the female lays 2 to 4 glossy white eggs in the bottom. Both sexes incubate the eggs and rear the young. The nest hole usually becomes very messy from the indigestible fruit pips regurgitated by the youngsters and parents.

Left: *Keel-breasted toucans inhabit warm lowland woodlands of tropical America. Their huge, brightly coloured bills are very light in construction. Little is known about why the bills of various toucans show such a range of shapes and colours.*

Perching birds

(ORDER PASSERIFORMES)

This huge order is a large, complex and highly developed one containing over 5,000 species. This is roughly three-quarters of the known bird species living today. Perching birds are characterized by their feet. All have 3 similar, unwebbed front toes and a hind toe which is highly developed but not reversible. None of the species reaches a very large size, the largest being the raven at 66 centimetres and the Australian lyrebirds, which are almost a metre long, including the tail.

Classification of the perching birds is very complex and varies from authority to authority. However there are about 56 families, although some say that there are as few as 50 and others as many as 70. The birds of this order are landbirds, only flying over the sea during migration. They have exploited all kinds of habitats, including deserts and mountains, as well as becoming adapted to live in man-made environments, from cities to agricultural lands, where they make use of 'unnatural' food. Passerines usually build nests, which range from the simple cup-shaped nests of thrushes and blackbirds to the intricate nests of the weaverbirds

and the stitched nests of tailorbirds. The young are always blind, naked and helpless at birth (nidicolous). Both sexes usually take part in caring for the young, although a few species, such as the 9 species of the widowbird family, are parasitic, laying their eggs in other birds' nests. Hosts' eggs usually remain in the nest.

Ovenbirds

The 221 species of ovenbird of tropical America are smallish birds that are very diverse in habit, although they are all rather drab in colour. The most interesting members are the true ovenbirds of the genus *Furnarius*

Left: *A section through the nest of a South American ovenbird showing the spiral corridor that leads to the nest chamber. Ovenbirds lay from 2 to 6 eggs, usually white. Both parents share the nesting duties, and the young hatch after 2 to 3 weeks. The nestlings remain in the nest from 12 to 18 days.*

since it is these that build large mud nests that are baked hard by the sun's rays. The widespread rufous ovenbird (*F. rufus*) of southern Brazil and northern Agentina has adapted to building its nest on fence posts or under the eaves of houses when suitable tree branches and stumps are not available. The mud oven has a hole through which the bird enters. The bird progresses along a spiral corridor into the nest chamber itself.

The 223 or more species of antbird, ant thrush, and ant pitta are closely related to the ovenbirds of Central and South America. They build simpler nests and their beaks are usually hooked at the tip.

An ovenbird and its nest

Ornamented cotingas

The cotinga family, which is composed of 90 or so species, contains some very showy and decorative birds. They are related to mana-kins and tyrants and live in Central and South America. The name cotinga is derived from an Indian word meaning 'white-washed' and refers to the white bellbird (*Procnias alba*). It is called a bellbird because of its ringing, far-carrying call. The most brilliant of cotingas are the cock-of-the-rocks. (*Rupicola* species). One species, from the Guianas and northern Brazil, is a fiery bright orange, while the Andean species is a brilliant red with black wings and tail. The male has on the head a fine erectile crest that forms a fan, hiding most of the beak.

The most bizarre cotinga is the umbrella bird. It is a black, crow-sized bird with an elaborate crest of curled feathers on its head (its umbrella) and a huge wattle hanging down from its chin.

Below: *The brilliant plumage of the Peruvian cock-of-the-rock* (Rupicola peruviana) *identifies the male sex. His female is dark brown. The peculiar crest is permanently erect and often covers the bill.*

Left: *A tiny male vermilion flycatcher waits for a fly to come within range. Then it will dart out, snap it up and return to its look-out perch. This species is one of the few brightly coloured tyrant flycatchers. The female is drabber.*

Tyrant flycatchers

The insectivorous tyrant fly-catchers hunt in open spaces, catching insects while in flight. The 365 species range through the Americas and are even found above the treeline in the moun-

Superb lyrebird male

Lyrebirds – Australian mimics

The majority of people know what a lyrebird looks like since it is widely familiar through its symbolic use on cards, stamps and seals originating from its Australian home. However, few people have seen one in the wild because it is a shy and secretive bird, inhabiting densely vegetated gullies in the temperate rainforests and eucalyptus forests of eastern Australia.

There are 2 species, the superb lyrebird (*Menura superba*) and the smaller Albert's lyrebird, this latter species being the more northerly one. It is the male bird that has given rise to the name. He is adorned with a long tail, which trails behind him when he is walking normally. However, during his courtship display, the tail feathers are thrown forward and raised over the head to form a shimmering lyre, the dark outer pair of feathers being lyre-shaped and framing the delicate filamentous, silvery feathers. The male displays from a dancing mound which he prepares from soil and leaves. He

has several mounds and visits each in turn. When an interested female is responsive he sings and displays in a frenzy until mating takes place, the female being covered by his 'lyre'.

The male probably mates with several females. Once mated each female goes off and builds a domed nest from moss and sticks, lining it with feathers from her back and legs. She lays only 1 egg and the chick hatches after 35 to 40 days. One unusual aspect of caring for the nest is shown by the female who carries all the droppings from the nest to the nearest stream.

tains. The majority of these birds are dull and drab, but usually the brown male has a colourful crest of feathers that he can raise. The male vermilion flycatcher (*Pyrocephalus rubinus*) however, is one species that is most vivid with its striking black and red plumage.

Larks

To be 'up with the lark' is the term given to early rising Europeans, and it relates to the fact that the lark is one of the first birds to sing in the spring and summer mornings. Among the 70 species of larks is the skylark (*Alauda arvensis*) that is probably the best known. The majority of larks are small brown birds, but many have crests adorning the head. They are widely distributed and have long, pointed wings, rounded, scaly ankles and long, straight, hind claws. Most of them live in open country and build their nest on the ground. They feed on insects, seeds, and other parts of plants in about equal amounts.

Swallows and martins

Swallows and martins are often confused with swifts, although they are quite distinct birds. Swifts are longer-winged and usually fly higher in the sky than swallows. Although they are all masters of flight, the swallows' flight is the more erratic. Unlike the swifts, swallows and martins often alight during the day on twigs, branches, wires and roofs. They do, however, spend most of the daylight hours flying back and forth seeking insects.

One of the most familiar and widespread of the family is the bird known in Britain as the swallow (*Hirundo rustica*). It is known to the French and Germans as the chimney swallow and to the Americans, Dutch and Norwegians as the barn swallow. The North American population migrates to South America to winter, while the European population flies to Africa, and the Asiatic population travels to

Malaya and the Philippines. This species, like most of its relatives, shows a strong faithfulness to its nesting site, returning year after year, often to the same nest. A pair work together bringing pellets of mud in their bills and adding straw and grass for strength. The 4 or 5 eggs are incubated mainly by the female who has a brood patch. This is a bare patch on her breast that is amply supplied with blood vessels to keep the eggs warm. The young hatch after 14 to 16 days and leave the nest about 3 weeks later, when they can fly. The house martin (*Delichon urbica*) of Eurasia also builds mud nests, but many swallows, such as the tree swallow of temperate North America, are cavity nesters. The purple martin of America will use bird-boxes, one kind being a gourd hung on a pole.

Below: *An attentive house martin parent flies in to feed its hungry offspring secure in the mud nest under the gutter of a house. Grass and roots add strength to the bird's home. Both parents help to feed the nestlings which fly at 20 days.*

Crows, magpies and jays

The crows, magpies and jays, of which there are some 102 species, can be summed up as noisy and aggressive birds, always bold and active in their habits. They are almost world-wide in distribution but are most common in the northern hemisphere. The crows are the more sombre of the family members, while the smaller jays are usually quite colourful.

The crows which are the larger-bodied birds of the perching-bird order, include the largest of them all, the raven (*Corvus corax*). This bird is the most widely distributed member of the family, being found practically all over the arctic and temperate northern hemisphere.

The smaller jays are very noisy most of the year but they become remarkably silent during the breeding season so as not to draw attention to their well-hidden nests, presumably. Like other members of their family, they are noted hoarders and will carry off glittering articles, from silver milk tops to rings. The widespread common jay (*Garrulus glandarius*) lives right across Europe to south-east Asia. It has a colourful patch of blue and black feathers at the bend of the wing. The blue jay (*Cyanacitta cristata*) of eastern North America has invaded city parks and gardens.

The magpie (*Pica pica*) is one of the few species that occurs both in Europe and North America. It is a bold bird, living solitarily or as one of a pair that stay together for life. In suburban gardens it will often chase off larger birds, squirrels, and even cats, that encroach within its territory.

Below: *A European jay is conspicuous not only for its colourful plumage but also for its screeching cry as it flies from tree to tree. The only time it is rather silent is during the nesting season. In autumn it buries acorns to eat later.*

Birds of paradise

The birds of paradise are some of the most beautiful and ornate members of the bird world. The males develop amazing plumage in order to attract the females. Usually they have long tail feathers and crests or ruffs. These birds acquired their name from the Spaniards who, in 1522, brought some of their feathers home from Magellan's first voyage around the world. The Spaniards thought them so magnificent that they believed that they must have come from paradise. Most of the 42 species live in New Guinea and its islands, but 4 are found in north-eastern Australia.

During courtship the gaudy males perform exciting displays which make the most of their magnificent plumage. The blue bird of paradise is a trapeze artist, hanging upside-down by his toes with his gold-tipped, blue plumes showering around him in a fine-spread fan of colour. The king bird of paradise is a dancer. He perches on a tree branch and opens his green plumes on either side of his white breast and scarlet body.

As with the related bowerbirds, the females depart after a successful mating to build their nest and rear their young alone.

Prince Rudolph's blue

Little king

Red
plumed

Bowerbirds

The bowerbirds live in the tropical forests of New Guinea and Australia. The males are brightly coloured and court the drab females by producing extravagant displays. Bowerbirds first build constructions of twigs and grass which they then decorate in various ways. These are called bowers because they resemble the garden bowers or small shelters that people sometimes build in their gardens. At the start of each annual breeding season the males either make new bowers or repair the old ones, and they may spend several weeks posing there to attract as many females as possible.

The 2 main groups of the 17 species of bowerbird are the maypole builders and the avenue builders. Maypole builders construct their bowers around the trunks of saplings. They are often called 'gardener birds' because they plant mosses in their dancing grounds. The crestless gardener (*Amblyornis inornatus*) of western New Guinea builds a cone-shaped hut that is sometimes over 2 metres high – quite a task for a 23 centimetre bird. The avenue

Above: *A male satin bowerbird attending to his bower, where he will court and mate with a female. She builds a nest elsewhere.*

builders make their bowers by first laying a thick mat of twigs. More twigs are added to raise low parallel walls on the mat.

Both groups decorate their buildings with colourful objects, including fruits, berries, flowers, insects, shells, bones and even man's rubbish, such as bottle tops and shiny foil.

The satin bowerbirds (*Ptilonorhynchus violaceus*) and the regent bowerbird (*Sericulus chrysocephalus*) also paint their bowers. They chew charcoal, grasses and fruit which mixed with their saliva makes a blue or greenish paint. They then take some leaves or bark, which they use as a brush, to daub the paint on the bowers.

After courtship and mating, the females leave the bower and depart to build their own nest and to raise their young alone. The drab colours of the females help hide and protect them and their young in the forest undergrowth.

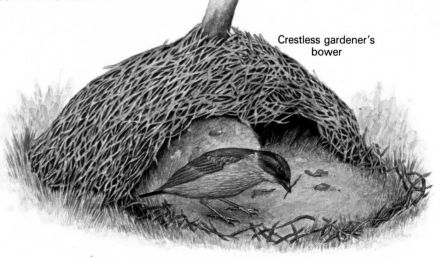

Crestless gardener's
bower

Tits, titmice and chickadees

The small tits, titmice and chickadees, of which there are some 65 species, show remarkable intelligence. In Britain, some great tits (*Parus major*) and blue tits (*P. caeruleus*) have learnt to peck the silver foil milk bottle tops to obtain the cream from the top of the milk. In Japan, the varied tit 'tells fortunes' at shrine festivals and street fairs by choosing a paper fortune and then tearing off the wrapping for the person. The majority of tits have grey or black, soft, thick plumage, with blue on the back and yellow on the breast. They are found throughout the world, except in South America, Australia and Madagascar, and are most numerous in temperate regions. They are continuously on the

The dipper – an aquatic perching bird

Very few members of the perching bird order readily take to water. The dipper (*Cinclus cinclus*) of Eurasia, however, loves cool mountain streams, running in and out of the water as it searches for small, aquatic invertebrates. It will even swim under the water, using its wings to propel it. It will also walk along the bottom of the stream bed, busily investigating it for food, but it can spend only about 10 seconds underwater before it must surface to breathe.

This bird loves the water so much that quite often the nest is built on a ledge or rock crevice behind a waterfall. Inside the domed nest of moss, the female incubates the 3 to 6 eggs and then after they have hatched both parents feed the young. After about 2 to 3 weeks the young leave the nest and they are as much at home in the water as their parents.

move, searching for insects cracks and holes in trees shrubs.

Most tits, titmice and chic dees nest in small holes in tr and several species will nes bird-boxes. They line their r with hair, fur or moss. The fem incubates the 4 to 12 eggs, once they have hatched both ents work very hard catch insects and grubs to feed hungry brood. The long-tailed

Wrens – tiny and busy birds

There are 60 species of wren. It is astonishing that these small birds should have such incredibly loud burbling voices. They are most numerous in South America but the winter wren (*Troglodytes troglodytes*) of Eurasia and North America is the species with which most people are familiar. It is the typical wren shape with a short, stubby body, short tail and pointed beak. Like most of the wren species it lives close to the ground, industriously searching the undergrowth for insects.

ove: *A blue tit stealing the cream. An
active garden bird of Europe, this tit
ws quite a lot of intelligence. It has
nt to peck open milk bottle tops to
ain food.*

egithalos caudatus) makes a
nest with a side entrance. It is
utifully woven from mosses,
hens and spider webs, and is
ed with masses of soft feathers.
e penduline tit (*Remiz pendu-
us*) makes a long, hanging bag-
ped nest.

Flycatchers, babblers, thrushes and warblers

These birds make up a huge family of about 1,200 species, most of which are found only in the Old World. The majority of the members have drab brown or grey plumage but some flycatchers are brightly coloured and have crests and long tails. In the main they are insectivorous, and so they feed chiefly on insects but will also eat spiders, snails and even some vegetable matter. The European song thrush (*Turdus philomelos*) displays an interesting feeding adaptation. It will use a stone as an 'anvil' on which to smash the shells of snails it is feeding on.

The thrushes include many familiar and well-loved birds, including the European robin (*Erithacus rubecula*). The related American robin is twice its size. The blackbirds, nightingales and bluebirds are also members of the thrush family.

The warblers, as their name indicates, are renowned for the quality of their song. The blackcap and garden warbler, for example, have lovely, ringing lively songs. Cuckoos often choose the nests of warblers, particularly the sedge and reed warblers, in which to lay their eggs. One of the strangest warbler nest is that built by the tailorbird of Asia. The cup-shaped nest is made by sewing together the edges of one or two leaves with either plant fibres or spiders' silk.

The African, Asian and Australian babblers are all noisy birds. Their loud chattering helps to keep the flocks together in their woodland habitat. Only species, the wren-tit, is found in the United States. Typical of this group is the cinna or quail-thrush of Australia, while the strangest members are the bald crow and bare-headed rock fowl of tropical West Africa. With a length of about 36 centimetres, they are also two of the larger members. The Cameroon species has a bare blue and red head, while the Sierra Leone one has a naked bright yellow head.

Starlings

The common starling (*Sturnus vulgaris*) is the best known of the 100 or so species which make up

Above left: *The nest of a tailorbird* (Orthotomus sutorius) *of Asia showing the leaves sewn together, forming a pocket. Inside this the nest proper is built.*
Left: *A tailorbird at its nest.*

Left: *A thrush's anvil. Each snail is brought to the stone and smashed so that the songbird can extract the soft body within. The bird will return again and again to his anvil with snails.*

the starling family, and it has been introduced by man to most parts of the world. These jaunty, aggressive birds are usually dark in colour with a metallic sheen. However, some are brightly coloured, for example the golden breasted starling or the superb starling, both of which are found in East Africa.

The Asiatic mynah birds are renowned for being among those birds which make the best talking cage-birds. The hill mynah and the common mynah are the specimens usually kept in captivity. Their imitation of human speech is much better than that achieved by parrots.

Mutual benefits between birds and other animals

Throughout the bird kingdom there are certain species that live in close association with another animal to the benefit of both animals. Two species of starling, the red-billed and yellow-billed oxpeckers of Africa, spend most of their days clinging to large mammals such as antelopes, cattle, giraffes, hippos and rhinos. They grip with robust claws and industriously seek out ticks, flies and other parasites that are living on the skin of the host mammal. The mammal benefits by being de-infested and the bird has a readily available source of food. At dusk the oxpeckers may fly off in flocks to roost among the reeds, but quite often they sleep on the host, which saves them having to find a new companion in the morning.

Some African birds are known to pick leeches and debris from the gums of basking crocodiles, and even to enter their open mouths. Usually such birds are Egyptian plovers although other plovers, dikkops, and sandpipers, are known to behave in this way. The birds obtain food and are of help to the reptiles in that they give early warning of any approaching danger.

Also in Africa, carmine bee-eaters often travel on the backs of kori bustards or grazing mammals. They fly down from their backs to pick up insects, such as locusts, that are disturbed by the host. Cattle egrets also pick up insects disturbed by grazing animals, but they usually feed at the feet of the grazers, although sometimes they will land on the backs of cattle or even elephants to gain a better vantage point.

Left: *Red-billed oxpeckers* (Buphagus species) *on their chosen domesticated cow. They stick with the animal like glue, using their long, sharp claws, taking ticks and other skin parasites as food.*

Honey-eaters, sunbirds and flowerpeckers

The honey-eaters, sunbirds and flowerpeckers form three closely related families. They are mostly small- to medium-sized birds which feed on nectar, berries and insects. They mainly inhabit the tropical forests of Asia, and Australia, although the brightly coloured sunbirds are also numerous in Africa and Madagascar.

The olive-green helmeted honey-eater of Australia is typical of the honey-eaters. Small flocks quarrel amongst the flowers of the gum trees in order to reach their favourite flowers. During its breeding season, the white-eared honey-eater will sometimes land on a person's head in order to pull out a few hairs with which to line its nest.

The brilliant sunbirds are often compared with the unrelated hummingbirds for their beauty, but they are not such expert fliers as the hummingbirds. In Africa their hanging purse-like nests are a common sight.

Flowerpeckers are the smallest birds that are commonly seen in the Oriental and Australian regions. They are active, plump little birds, and true to their name, spend lots of their time pecking flowers for nectar or insects. One Australian species, the mistletoe bird, pecks tropical mistletoe and is largely responsible for the spread of this parasite.

Sparrows, finches and weavers – small seed-eaters

Finches are small, tree-loving seed-eaters that are widely distributed throughout most of the world. Many of these birds migrate to warmer climates in winter.

Above: *A honeyeater perches on the ripe fruit of an umbrella tree in a tropical fores of Australia.*

The main characteristic of sparrows and finches is their short, conical, pointed bill, which is adapted to eating seeds. There are over 315 species in the New World, including the colourfu cardinals and the sparrows (Europeans call them buntings). One particularly interesting group is that comprising the Darwin's fin ches of the Galapagos Islands From a single parent stock that reached the islands, some 14 species have evolved, each adapted to eat a slightly different food One species, the woodpecke finch, uses a twig or cactus spine to extract insects from crevices and holes.

In the Old World there are some 375 or so species of seed-eater including the goldfinches, the waxbills and weaver-birds. Many of the tropical species are beauti

Above: *Colonial suspended nests of the weaver bird, each one skilfully woven of plant fibres with the bill and feet.*
Left: *A weaver perched by its nest. The suspended home gives protection from predators such as tree-dwelling snakes.*

ful and, unfortunately, are highly prized as cage-birds. The Gouldian finch of northern Australia, with its purple and yellow breast, is one such species.

The weaverbirds or weaver finches include the familiar house sparrow. The remarkable feature of this group is the ability of many of the species to weave highly complex nests. With the bill and feet they make these nests from plant fibres, shaping them into long, hanging pouches, often with a tubular entrance hole. From a distance, a tree with several nests in it looks as if it is adorned with fruits or decorations. The sociable weaverbirds have gone a stage further, building huge, thatched, communal nests which may house hundreds of birds.

Mammals (CLASS MAMMALIA)

Mammals have characteristi
features which distinguish then
from the other vertebrates. The
all have hair, although it may b
reduced to a few bristles or tufts
as in rhinoceroses and whales. A
mammals feed their young o
milk produced in the specia
mammary glands of the mother
Mammalian young are born alive
except for one order, the mono
tremes, which have retained thei
reptilian ancestors' habit of lay

Wildebeest move steadily onwards on their annual migration cycle on the Serengeti Plain of East Africa in search of fresh green grass and water. They reach each area when the forage is in peak condition.

ing eggs. Mammals also have a heart that is divided into four chambers, and which allows a high metabolic rate and an active life. Mammals can produce heat to maintain a constant body temperature. Together with the birds, mammals are the only warm-blooded animals in the world.

Although there are only some 4,200 species of mammal, they have evolved, during their time on the Earth, to fill many available niches. The air is occupied by the 900 or more species of bat, as well as being used by the gliding squirrels and phalangers. Trees are the home of many mammals such as squirrels, monkeys and apes. On the ground, cattle, antelopes and horses graze the open grasslands, while forests are the home of the browsing deer. Leaf litter is the hunting place for insectivorous shrews as well as many small herbivorous rodents. Beneath the ground are burrowing moles and certain rodents such as chipmunks and prairie dogs. Many mammals, such as badgers and armadillos, retreat underground to sleep. Most of these land mammals mentioned are preyed upon by the active flesh-eating mammals which include the wild cats, wild dogs, weasels and martins. Mammals have also returned to the sea,

whales, dolphins, porpoises and sea-cows even giving birth at sea. The seals and their relatives, however, are still obliged to return to land to mate and produce young. Mammals can also live in extreme climates, the polar bear being adapted to Arctic conditions, sheep and goats to the high mountains and gerbils and fennec foxes to desert life. Although not large in number of species, the mammals are certainly highly successful and much is known about their habits and behaviour. One reason for this is that man himself is a mammal and this, no doubt, has led him to find out as much as possible about his relatives during the course of his history.

Mammals originated from primitive reptiles about 200 million years ago, and fossils show that they looked very like the living shrews. At this time the dinosaurs and other reptiles dominated the land. For about 100 million years during the Age of Reptiles, the mammals remained small, and probably nocturnal, hunting for insects to eat at night. They probably competed with small lizards for their food. During Jurassic times there were 5 well established orders of mammal. Three of these orders were to die out but it is thought that the other 2 were the ancestors of all modern mammals. The docodonts possibly gave rise to the living monotremes – the platypus and spiny anteaters. The pantotheres probably gave rise to the pouched marsupials and the placental mammals. Placental mammals form the large group that has the baby growing as an embryo in the womb of the mother until it is mature enough to be born and survive as an individual.

When the dinosaurs became extinct about 70 million years ago, the mammals survived, and they evolved new species quite quickly to become the new dominant land animals. One group was carnivorous (flesh-eating), and contained mammals that looked very similar to modern cats and dogs. There were also several groups of herbivores (plant-eaters) at this time. One was the American *Uintatherium*, which reached 4 metres in length. It was built like a modern rhinoceros, had sabre-like tusks, and three pairs of horny swellings on the top of its skull.

Between 35 million and 20 million years ago the ancestors of most living mammals were evolving. Giant rhinos, such as *Paraceatherium* which reached 8 metres in length, roamed the Ameri-

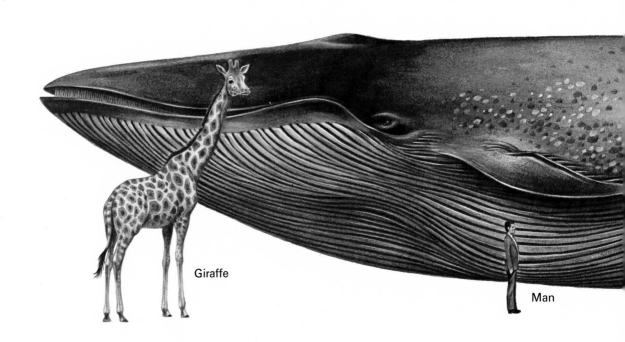

Giraffe

Man

Facts and figures in the world of mammals

There is a quite wide and dramatic range in the size of mammals. The smallest species is the Etruscan or Savi's pygmy shrew at 1.5 grams, while at the other end is the gigantic blue whale that tips the scales at 150,000 kilograms. The back of this cetacean would take some 10 African elephants standing head to tail along its length. The African elephant is the largest land mammal at 4-6 tonnes.

The tallest living animal is a mammal, the African giraffe. A Mosai bull measured 6.09 metres, and this is 1.78 metres taller than a double-decker bus. The fastest land animal is the cheetah which has a probable maximum speed of 96 to 101 kilometres per hour over a short distance.

can landscape. This animal was as tall as the living giraffe and it browsed on the leaves of tall trees. In North America early cats and horses were also evolving. The first horse *Hyracotherium*, often referred to as *Eohippus*, is known from fossils from rocks 50 million years old. This horse was about the size of a modern fox terrier. It lived in dense tropical forests, but later forms evolved to live on the plains of North America and these gradually evolved into species that reached other parts of the world. In Africa 30 million years ago there were rhinoceroses, early elephants and hyraxes on land, with whales and sea-cows in the water.

The South American continent was cut off from North America during the Tertiary period and it was here that several groups of mammals were able to evolve without competition from flesh-eaters. The marsupial opossums and New World monkeys radiated into many different species. The order Edentata which includes the armadillos, anteaters and sloths evolved many giant forms. The giant ground sloth (*Megatherium*) was over 6 metres long and weighed several tonnes. It lived during the Pleistocene epoch and early man may well have seen this mammal when he invaded the continent. The cousin to modern armadillos, *Glyptodon*, also lived about this time. Its bony armour was fused into a complete tortoise-like shell.

Plains and grasslands stretched right across Europe to China some 10 million years ago and gazelles, antelopes, rhinoceroses and giraffes grazed on them. They were preyed on by big cats and hyenas. Then about 1 million years ago the Ice Ages began. This had a dramatic effect on the mammals, as also did man, who was becoming an important hunter at this time. Why the Ice Ages occurred has yet to be explained, but during the last million years the Arctic ice expanded to reach southwards at least three times. In cold periods, mammoths, polar bears, reindeer, wolves and woolly rhinoceroses roamed southern Europe. In warm periods animals such as sabre-toothed cats, giant deer, elephants, cave lions and horses invaded the area. Many of these forms became extinct in later Ice Ages, although the reason for this is unknown. Those that became extinct were mainly the larger forms and the surviving mammals gave rise to the mammals that we find living on the Earth today.

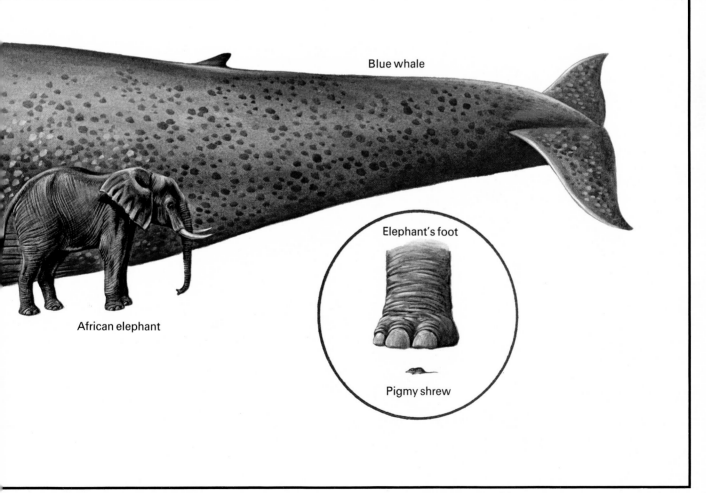

Blue whale

African elephant

Elephant's foot

Pigmy shrew

Monotremes – the egg-laying mammals

(ORDER MONOTREMATA)

The monotremes of the order Monotremata are the most primitive of all the mammals and little is known of their history. The only known fossils have been discovered in Australia and the living species are confined to Australia, Tasmania and New Guinea. The 6 species are of two basic types, the platypus (*Ornithorhynchus anatinus*) being one, and the 5 species of echidna (*Tachyglossus* and *Zaglossus* species) being the other.

The platypus is placed in its own family, the Ornithorhynchidae. It inhabits streams and lakes in eastern Australia and Tasmania. The male is slightly larger than the female, measuring about 56 centimetres in length and around 2 kilograms in weight. The short, dense, soft fur is waterproof, which is extremely important to this aquatic animal. It swims very skilfully, using its flattened tail as a rudder and obtaining propulsion from its webbed feet. These creatures are active mainly in the morning and evening, silently searching for food such as crayfish, shrimps, larvae of water insects, snails, tadpoles, worms and small fishes. Captive platypuses are known to eat about half their weight a day.

Platypuses mate from August to October in water, after the male and female have swum in circles and played about in an unusual and intricate courtship. The female carries wet leaves into a deep and elaborate nest burrow and lays 1 to 3 eggs on this bedding. The wet leaves prevent the small, rubbery eggs from be-

Below: The curious duck-billed platypus lives in slow-flowing streams of Australia. The bill filters small aquatic animals.

Right: *An echidna or spiny anteater. Its long snout shows that this species comes from New Guinea and not Australia.*

coming too dry. The mother curls her body round them after plugging up the den's entrance. She leaves the den every few days to wet her hair and defecate.

When the eggs hatch the youngsters are blind and naked, and about 2.5 centimetres long. They are fed by milk which flows from pores opening on the mother's belly. The young instinctively lap it up. They do not venture from the burrow until they are some 16 weeks old, fully furred and about 33 centimetres long. During this time the male uses a shelter burrow nearby, never being allowed near his female after courtship and mating.

The echidnas or spiny anteaters are completely different in appearance, with the furry body covered with sharp spines, formed from modified hairs. They are rather round creatures with short legs and broad feet equipped with strong claws for digging. The jaws are long and narrow, ending in a short snout in the 2 *Tachyglossus* species and a very long snout in the 3 *Zaglossus* species. A long, sticky tongue capturing ants, other insects and worms, is shot in and out of the snout.

In the breeding season a female echidna develops a temporary pouch into which the single egg (occasionally 2) is transferred as soon as it is laid. How this is done

is not known, but the egg hatches after 7 to 10 days into a naked, blind baby, 1.2 centimetres long. It laps the yellowish milk that flows along several hair tufts from the ducts of the milk glands within the pouch. The youngster grows inside the pouch until its spines begin to develop between the age of 6 and 8 weeks. It is about 10 centimetres long now,

and the mother deposits it in a sheltered spot. It is not known whether she continues to feed it or whether she leaves it to its own devices.

Below: *The platypus hides most of the day in a complex series of burrows that lead from its stream. The female lays her two eggs in one burrow lined with moist leaves. On hatching, the young lap milk from her belly.*

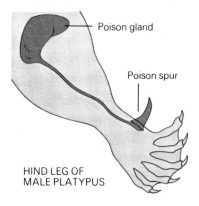

Poison gland

Poison spur

HIND LEG OF MALE PLATYPUS

Marsupials – the pouched mammals

(ORDER MARSUPIALIA)

The marsupials are the pouched mammals, and like the monotremes, they are more primitive than the other mammalian orders. The name marsupial means 'pouched-one' and this is because the female members of the order usually have a protective pouch of skin around their nipples. The female does not lay eggs but her offspring are born at such an early stage of development that they look like embryos. However, the young are always strong enough to crawl into the mother's pouch from the birth canal or cloaca, even though they are blind and naked. The young marsupial stays inside the pouch attached to a teat, feeding and growing for several weeks, the exact length of time depending on the species. At last it is sufficiently developed to take a look at the outside world and it begins to explore its mother's body outside the pouch, but retreats inside if danger threatens.

On the whole, marsupials are small-brained creatures with only

Above: *A common or Virginian opossum with her growing litter. When born these were tiny and made their way into her pouch. These babies are weaned and so are carried on their mother's back, clinging to the fur.*

little intelligence, and during their evolution they lost the battle in competition with the quicker, more efficient placental mammals. Today they are found only in two areas, South and Central America, and Australasia. Their survival is due to these continents being isolated from the parts of the world where the other mammalian orders were evolving at a rapid rate.

American opossums

The only marsupials living in America are the opossums, of which there are some 65 species. The best known and most successful is the Virginian or common opossum (*Didelphis virginiana*), which was once confined to South America. However, over thousands of years, it flourished and invaded North America, where it is still spreading. It is the only marsupial found in North America. It owes its success to its adaptability, varied diet, rapid breeding rate and habit of playing possum. The common opossum looks rather like a long-haired rat with a prehensile, naked tail, although it is the size of a domestic cat. The female gives birth after only 12 days gestation to pea-sized babies weighing about 2 grams. Although 20 babies may be born, only about a dozen survive as the mother has only 12 or 13 nipples inside her pouch. The youngsters do not emerge for 100 days, and then they cling to the mother's back until they become too large.

The other species of opossum either look like the common opossum and are a similar size, or they are much smaller, being mouse or rat-sized.

Most opossums are nocturnal, searching most of the night for all kinds of food, which ranges from small animals, insects and carrion to fruit and grain. They lead rather squirrel-like lives, although the yapok or water opossum (*Chironectes minimus*) lives an otter-style life, spending much time in the water. It has webbed feet to help it move through the water and it rests and breeds in burrows in the river bank. It feeds on shrimps and small fishes.

Playing possum

The opossum's behaviour has given rise to the saying 'playing possum' when a human or indeed another animal feigns death. The marsupial, when endangered by an attacker, lies limply on its side, its tongue hanging out of its mouth and its eyes shut. In this state it allows itself to be mauled about and played with until the attacker loses interest and departs. Other animals that use this technique to escape predators include snakes, small mammals such as rodents, and lizards.

Australian pouched mammals

In the Australasian region, the marsupials have evolved over a long period of time to give a wide range of pouched mammals adapted to the various habitats and niches. There are species which are very similar to placental mammalian species in appearance and adaptations, and yet they are pouched animals. Thus there are marsupial cats, mice, rats, wolves, and moles.

The several species of marsupial native cats are better known as dasyures or quolls, the largest species (*Dasyurus quoll*), in fact, being cat-sized. The others all look like small placental carnivores, which is not surprising since the dasyures are small flesh-eaters. A fierce relative is the Tasmanian devil (*Sarcophilus harrisii*), which is now the largest carnivorous marsupial on the continent, as it seems that the thylacine is probably extinct. The 'devil' has certainly been persecuted by settlers, as it made a nuisance of itself stealing chickens. It is in no

immediate danger of extinction, although it has retreated to the more remote and rocky parts of the large island. It probably looks far more ferocious than it really is. It has enormous jaws which it opens wide when frightened, at the same time giving a wheezing snarl. The heavily built body is supported on rather short legs. It does have a voracious appetite and feeds on a variety of animals including rat-kangaroos, wallabies, lizards and birds such as parrots and quail.

Above: *Western Australian native cat (Satanellus hallucatus) is a small hunter of gum forest and grass savanna with scattered trees. Its diet is made up of small mammals, birds, lizards, fish and insects.*

The Tasmanian devil is closely related to the thylacine or Tasmanian wolf (*Thylacinus cynocephalus*), which is dog-like and 1.5 metres long. Its grey-brown back has 17 dark vertical stripes. This appearance has also led to its other nickname – 'tiger'. However, it is very probable that this marsupial tiger is now extinct, having been exterminated on the Australian mainland by settlers and the dingo, and in Tasmania by settlers. No living specimen has been seen for years and all recent sightings were probably domestic dogs.

The marsupial mice are similar in appearance and size to the house mouse. Their noses are usually more pointed, however, and the tail is shorter and fatter. They are mostly carnivorous, however, unlike the omnivorous rodents. They are attractive little animals with small pouches.

Left: *A Tasmanian devil, although a marsupial, resembles a small sun bear. It feeds on a variety of animals including poisonous black tiger snakes, sheep and chickens, as well as carcasses and bones.*

Right: *A marsupial mouse on the alert. There are several species of these pouched mammals, most of them being common in many areas, but not in settled regions where the introduced house cat preys on them.*

Quite often there are as many as 10 young, and they will cling to the mother's sides and back when they leave the pouch. Certain species, such as the crest-tailed marsupial rat (*Dasyuroides byrnei*), are as large as a rat, while others look more like the jumping jerboas.

The bandicoots of Australia look rather like large shrews. Their name is a corruption of an Indian word meaning pig-rat and it is thought that the early explorer Bass used this name first, in about 1799. The 19 species inhabit open plains, thick grass along the banks of swamps and rivers, thick scrub and forests. Members of the genus *Perameles* are found near towns and cities and frequently annoy gardeners by digging holes in gardens and lawns in their search for insects.

Female bandicoots usually have 8 mammae, although 6 or 10 are not uncommon. These numbers are quite sufficient to cater for the 2 to 6 young born in a litter. The pouch opens downwards and backwards and the newly-born young have tiny claws which are shed soon after they have been used to help the animal reach the pouch.

Although bandicoots are protected in Australia by the Fauna Preservation Act, several species are in danger of extinction, mainly because the Australian aborigines hunt them for food.

The family group of the phalangers includes mouse-like, squirrel-like, and lemur-like forms and the teddy-bear form, the koala. It also includes the Australian opossums, the name which Captain Cook gave to the ring-tails (*Pseudocheirus*) at Cooktown, in 1770, because of their superficial resemblance to the American opossums. Today the Australian ones are often called 'possums' to distinguish them. The more suitable name phalanger refers to the adaptation of some of the phalanges (finger and toe bones) to aid in climbing.

The cuscuses are the largest of the family, the spotted cuscus (*Phalanger maculatus*) being about as large as a domestic cat but more elongated. The sexes of this species are very distinct, the male being covered with large, irregular, pale blotches, while the female is a plain colour. The young go through a sequence of

Below: *A rabbit-eared bandicoot* (Macrotis lagotis) *demonstrates its burrowing skills. The snout forages for insect larvae in sandy soil, leaving conical mounds in the earth.*

Flying marsupials

There are 5 species of flying phalanger or 'gliders' as they are called, due to the fact that they take great leaps from tree to tree, using their gliding membrane. This membrane is a web of loose skin that runs between the front and hind limbs. The fluffy tail is used for steering when the animal is in the air, distances up to 90 metres being covered. The largest of these marsupials is the greater gliding possum (*Schoinobates volans*), a very attractive animal with soft and silky fur. During the day these animals shelter, either singly or in pairs, in hollows high up in trees. They rarely come down to the ground but occasionally will journey across open ground to another tree. During the breeding season their loud gurgling shrieks pierce the air.

Sugar glider
(*Petaurus breviceps*)

Below: Portrait of a spotted cuscus. This heavy, rather powerfully built phalanger is a nocturnal, tree-dwelling marsupial with a strong, prehensile tail. Here it shows how it rests during the day, curled up in the fork of a tree. It moves slowly and sluggishly, searching for insects, birds' eggs, and birds when it is quick enough to catch them. A female usually gives birth to 1 or 2 young.

colour changes as they grow. Cuscuses are tree-dwelling animals, moving around slowly at night, surprising small roosting birds and some lizards, which they eat. However, most of their food consists of leaves and fruit.

The brush-tailed phalanger (*Trichosurus vulpecula*) is the most widely distributed Australian marsupial, being quite plentiful over most of its range. This is despite the fact that it is hunted and killed for its fur, which is sold as 'Adelaide Chinchilla'. It is one of the few phal-

ngers to have a long, bushy,
prehensile tail.

Wombats

The hairy, chunky wombat (*Vombatus hirsutus*) is very similar in appearance to the koala, but it is a powerful digger rather than a climber. It is just under a metre long and lives on the rough hillsides of coastal, south-eastern Australia and Tasmania, where it moves about slowly with a wobbling gait. It is the marsupial bulldozer, digging extensive burrows, 30 metres long. Its chisel-shaped rodent-like front teeth grow continually but are worn down to a reasonable length by the animal eating bark, roots and leaves. The wombat lives alone, except during the breeding season. Like the koala the female has one offspring a year. The other surviving wombat species is the softer-haired hairy-nosed wombat (*Lasiorhinus latifrons*). It is interesting to note that fossil evidence has shown that there once was a wombat species as large as a hippopotamus.

Below: A wombat, caught away from its burrow which it digs rapidly with powerful clawed front feet.

The koala

The most loved of the phalanger family and perhaps the whole order of marsupials, is the koala (*Phascolarctos cinereus*), with its large head, big hairy ears, and large nose producing an appealing yet comical look. The koala is confined to eucalyptus forests in eastern Australia, as the leaves and young bark of these 'gum' trees make up most of its diet. The dense, woolly fur is greyish above and whitish below, and there is only a vestigial tail. Cheek pouches and a large caecum aid in dealing with the specialized diet, over a kilogram of leaves being eaten every day. The name koala means 'no drink' and indeed the animal does not do so.

The female bears only 1 young at a time. It stays in her backward opening pouch for 6 months, after which time it climbs on her back and travels piggy-back style. Because their habitat has been destroyed for agriculture and towns, and they have been slaughtered for their warm, durable, beautiful fur, numbers have been drastically reduced during this century. However, they are now protected and are being re-established in certain areas, such as Victoria. Thankfully the koala is in no danger of extinction.

grey kangaroos (*M. giganteus*) are the giants of the 'roo' family. An adult male or 'boomer' stands 2 metres high, and when moving at a slow pace he covers between 1.2 and 1.9 metres in a leap. A leap covers 9 metres or more, when he is travelling at 48 kilometres per hour over a short distance of open country. Although the two large species are called red and grey, various shades of red, brown, grey or black predominate. Males of the red kangaroo are usually rufous while the females are bluish grey and often called 'blue fliers'. Although it is difficult to tell the species at a distance, in the wild the red inhabits the open plains, while the grey prefers the open forests. The third species of kangaroo, the wallaroo (*M. robustus*) lives in rocky, hilly areas.

Wallabies

Wallabies are very similar in habit to kangaroos but are about half their size. The hare wallabies, as their name suggests, are rather hare-like but slightly larger than European hares. They resemble them in their movement and to some extent in habit. They usually live alone and rest by day

Kangaroos

There are some 55 species in the family of kangaroos and wallabies, which vary in size from the small musky rat-kangaroo (*Hypsiprymnodon moschatus*) of rainforest Queensland, which reaches 33 centimetres in length, to the red and grey kangaroos of the genus *Macropus*, which reach 2 metres in height. All are jumpers, the large, powerful hind legs propelling them along in leaps and bounds with the long, thick, tail being used as a prop, balancing organ or extra leg. Most members are nocturnal, resting in grassy 'nests' during the day. Sometimes they sunbathe on warm afternoons, but they become most active at dusk, searching for food until dawn. Most species graze or browse on the vegetation, but the rat-kangaroos eat animals, such as insects as well.

The red (*Macropus rufus*) and

Kangaroo birth

There is no definite breeding season in the kangaroo's life but most of the young are born in the winter. After a gestation period of between 30 and 40 days, a single joey — a baby kangaroo — is born, weighing under a gram and being less than 2 centimetres long. Although blind and naked, this tiny creature has quite well developed forelimbs and crawls up to the mother's pouch along a path the female has licked on her belly fur. Here it selects 1 of the 4 teats and becomes firmly attached to it for the next 33 weeks until it is completely developed. It now weighs about 3.5 kilograms and begins to venture out of the safety of the pouch, but whenever it senses

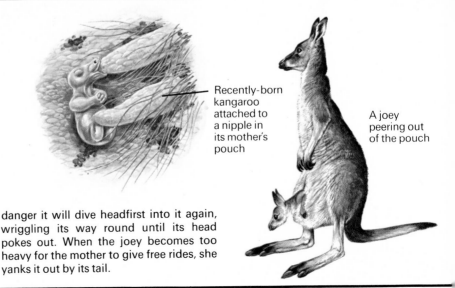

Recently-born kangaroo attached to a nipple in its mother's pouch

A joey peering out of the pouch

danger it will dive headfirst into it again, wriggling its way round until its head pokes out. When the joey becomes too heavy for the mother to give free rides, she yanks it out by its tail.

in a 'form', scratched in the ground under the shade of a bush or tufts of grasses. At night they nibble the vegetation. The rock wallabies (*Petrogale* species) usually inhabit rocky ranges and boulder-strewn outcrops. They are most common in some areas of central Australia. Their agility among the rocks is astonishing, some being able to leap up to 4 metres horizontally. Scrub wallabies (*Thylogale* species) inhabit swampy areas with thick scrub, grasses and fern tangles, as well as the undercover of heavy forests. They have been hunted and trapped for their meat and fur, and so, over a large part of their range, they are much reduced in numbers.

A surprise to many people, is the fact that there are some 5 species of kangaroos in New Guinea and 2 species in north-eastern Queensland that spend most of their time in trees. Tree kangaroos (*Dendrolagus* species) are very agile, often travelling rapidly from tree to tree, and leaping as much as 9 metres from one branch to a lower one. Unlike those of their ground relatives, the hind and forelimbs are of nearly equal length. The pads on the feet are covered with non-slip, roughened skin and some of their nails are curved to assist in gripping. The long, furred tail is the same thickness down its entire length, and, although not prehensile, it acts as a balancing organ. Tree kangaroos are not rare, but as they inhabit rather inaccessible areas their habits are not well known. They do come down to the ground and usually associate in small groups, feeding on fruit and leaves.

Above, right: *Ring-tailed rock wallabies* (Petrogale xanthopus) *emerge at dusk from their day-time retreat.*
Right: *Tree-kangaroos are found in the tropical rain forests of northern Queensland and New Guinea. The feet have strong curved claws for gripping.*

Insectivores – the insect-eaters

(ORDER INSECTIVORA)

The order Insectivora contains more than 370 species, over two thirds of these being various kinds of shrew. The insectivores are difficult to describe as a group because they lack any special, advanced characters and are broadly-speaking primitive, small-bodied creatures that feed largely on invertebrates such as worms, snails, insects and crustaceans. As well as the shrews, the group contains the hedgehogs, tenrecs, solenodons, and moles.

In most species the eyes are poorly developed, the brain is small and the teeth are rather unspecialized, being sharp and pointed. The fingers and toes are always clawed and an insectivore does not have opposable digits, so that it is incapable of grasping. The order is a living reminder of the type of animals that were the ancestors of more advanced groups. They are found today throughout Europe, Asia, Africa, North America and Central America, as well as Madagascar and the West Indies, with a small number in South America.

The solenodons (*Solenodon*) of the West Indies are rat-sized with long, naked tails, and long, greatly elongated snouts that have nostrils at the tip. The eyes are very small and the sense of smell is the most important factor in both finding food and their way about. Scent-producing musk glands are found in the armpits and groin regions and, probably because of this, the female's nipples are on her buttocks. Solenodons are becoming quite rare, since many are killed by the dogs and mongooses which have been introduced to the islands, and since the female produces only 1 to 3 young in each litter, the numbers are not made up.

Tenrecs are another island species, being found only on the

Below: *A European hedgehog mother and her family. The spines protect these insectivores from enemies such as foxes and badgers, especially when they roll tightly into a ball so the vulnerable belly is hidden.*

island of Madagascar. The 30 species are all quite small and have become adapted to various life-styles. Some are swimmers, others are climbers with long tails, and some live mole-like lives in underground burrows. Some, such as the hedgehog tenrec (*Setifer setosus*), are prickly, the hair having evolved into sharp spines. These insectivores give birth to many young. The common tenrec (*Tenrec ecaudatus*) is recorded to have produced a litter of 21.

Although only some tenrecs have spines, all 16 species of hedgehog have this excellent protection against predators. Many people are amazed at the noise a hedgehog makes as it snorts and grunts on its nightly feeding trips. Because it has no fear of predators, it is able to move about so noisily and boldly and when asleep it snores loudly! The hedgehogs of Europe and Asia need to hibernate through the cold winter months, surviving on large stores of fat built up during the fertile summer months. The European hedgehog (*Erinaceus europaeus*) is a friend of farmer and gardener, as it eats many pests such as insects, slugs, and snails, as well as small animals, worms and some vegetable matter.

The moon-rats are also known as hairy hedgehogs or gymnures. They are related to the hedgehogs, but lack spines. The 4 species are all found in south-east Asia. The moon-rat (*Echinosorex gymnurus*) is the largest living insectivore, being about 40 centimetres long plus a 20 centimetre tail.

Most shrews are small, mouse-like mammals with long, pointed snouts and short legs. They are very active and have to eat their own body weight in food every 24 hours in order to sustain their high metabolic rate. They are also extremely nervous and the

heart of a shrew may beat 20 times a second. It is thus not surprising that some die when frightened – by a loud clap of thunder for instance. The ferocity that these little animals show is quite amazing. They live alone, but if one encounters another on a hunting trip, then they attack each other, wrestling in a ball or on hind legs, squealing shrilly all the time. However, they are not as brave if a predator, such as a bird of prey, comes within their territory of

Above: *South-east Asian moon-rat snarls at being disturbed. This nocturnal 'hairy hedgehog' has no spines and is said to smell like rotten onions or sweaty feet, due to odour released from anal glands.*

about 10 square metres. Then they dive for their nearest retreat. Many shrews have salivary glands that can secrete poisonous substances that quickly paralyse small prey such as worms or slugs. If a human is bitten, the poison causes great pain.

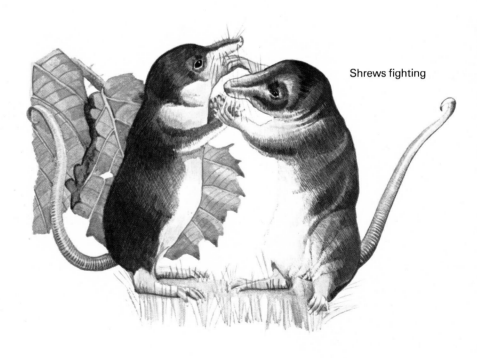

Shrews fighting

The world's smallest mammal

The smallest mammal in the world is a shrew, the Etruscan shrew (*Suncus etruscus*) which when fully grown is only 3.6 to 5.2 centimetres from the head to the end of its body, with a tail length of about 2.5 centimetres. It is so small that tunnels made by large earthworms offer it good passages and escape routes.

Another interesting aspect of the behaviour of certain shrew species is the habit of their young in holding on to each other and their mother in early exploration. With their jaws they take hold of the rump of the animal in front. One species which shows this fascinating aspect of behaviour is the white-toothed shrew (*Crocidura leucodon*).

A female shrew has her first litter when a year old, giving birth to between 2 and 10 blind, naked young in a nest of dried leaves and grasses. The young are independent within a month, and in fact need to be since the mother dies with the approach of winter when she is usually no older than 15 months.

The 20 or so species of mole are the diggers of the shrew world. There are four mole species living over most of Eurasia and Japan, but they cannot live in the Arctic or the central Asian Plateau since the frozen soil is too compact. America also has several mole species, the most curious being the star-nosed mole (*Condylura cristata*). Its distinctive feature, as the common name suggests, is the peculiar muzzle which is ringed with 22 pink flesh 'tentacles'. This is probably an adaptation for underground life, giving the mole a greater sensitivity to its surroundings. The star 'nose' is even conspicuous on the new-born moles.

Most moles have dark, dense soft and velvety fur which lie without direction, an adaptation which allows them to go backwards and forwards in their tight burrows without friction. Out of their burrows, moles find it difficult to walk on their strong, well clawed feet, and clumsily 'row' themselves along. Below ground however, the broad forepaws loos

Below: *Digger of the shrew world, the mole. Broad-forepaws with strong straight claws loosen the soil as the mole burrows.*

White-toothed shrews in procession

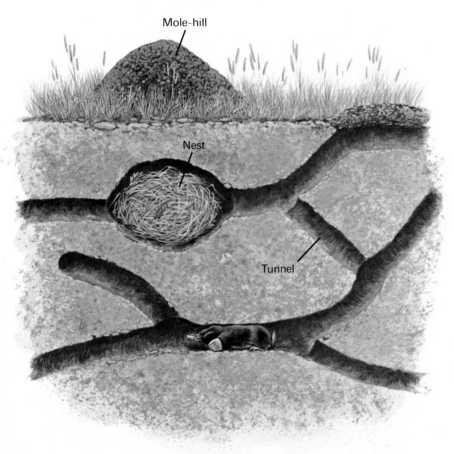

Mole-hill

Nest

Tunnel

en the soil and push it back to the strong hind feet which then shovel it out to the surface. Since they live entirely in the dark, the eyes of moles are tiny, less than 0.1 centimetre in diameter. External ear flaps would also get in the way and so none protrude. There are ear openings, however, at the back of the head, enabling the mole to hear quite well.

Moles are usually solitary, although they may share a network of burrows. They are active day and night, searching for worms and insect larvae and other small invertebrates, as they travel or dig through their burrows. Mole-hills are the result of earth being pushed up by the mole from deeper tunnels, the surface tunnels showing as surface ridges of soil. The deep tunnels are used for shelter and rearing the young. The nest, which is usually situated under a mound, is called a 'fortress'. Here, after about a month's gestation, between 1 and 7 young are born.

Moving moles

Although the tunnelling of moles annoys gardeners when it upturns a beautiful lawn, it is of great value in maintaining and developing soil. These attractive creatures can be prevented from damaging a piece of ground without resort to trapping or poisoning them. Apparently if empty glass bottles are put at an angle to the surface of the ground with the bottoms in the mole burrow and the necks sticking out, when the wind blows the piping, whistling sound drives the moles away from that ground.

Bats – the flying mammals

(ORDER CHIROPTERA)

Bats are the only mammals capable of true flight, and they are amongst the most successful of all mammals. There are almost 1,000 species, making this order second only in size to the rodent order. Bats are found living in almost all parts of the world, except the polar regions. Oceanic islands have been invaded and even New Zealand has 2 species of native bat.

Bats show much in common with the insectivores as regards basic structure, and it is accepted that the two orders had a common ancestor. During their evolution bats probably passed through a gliding stage and then became powerful fliers, rivalling birds in their expertise. As they evolved, they diverged in two main directions, leading to the fruit eaters and the insect eaters, but the body structure of the two groups is similar. Their forelimbs have become greatly elongated, especially the 4 fingers which play such an important part in the support of the wing membranes. The thumb acts as a free finger, and the bat uses this to hang upside down or to climb about in trees. The skin stretches from the fingers backwards supported by the sides of the body, the back legs and the tail.

Left: *Long-eared bat* (Plecotus *species*) *in flight showing how the front arms and fingers have elongated to allow skin to be stretched between them and the hind legs and tail. This bat catches insects in flight.*

Above: *A flying fox at rest, hanging by its feet with its wings folded round its chest. This fruit bat sucks the ripe flesh of fruit and is most unpopular in south-east Asia and Australia with tropical fruit growers.*

Megachiropterans or fruit bats

The fruit bats of the sub-order Megachiroptera differ from the insectivorous bats in being far less specialized for night flying. Only a minority of the 150 or more species can navigate by echolocation – the remainder are completely helpless if blindfolded, since they rely on their large eyes for finding their way around. The fruit bats are all confined to the Old World. The family includes the flying foxes, the 51 species all being found in the Malaysian Pacific region. The largest bat is a flying fox (*Pteropus giganteus*), which may have a wingspan of up to 1.7 metres although it weighs only about 1 kilogram. The flying foxes are not related to real foxes but the large, dog-like muzzle, large ears and round, efficient eyes make the head look rather like that of a fox. The strangest member of the fruit bats, is an African tropical species, the hammer-headed bat (*Hypsignathus monstrosus*), which has an almost horse-like head.

Fruit bats as a rule eat fruits such as dates, figs, bananas, pau pau and guavas. The bat takes the fruit into its mouth and then the pointed canines and premolars pierce the skin so that the ripe flesh can be sucked and crushed until only a dryish pulp is left. It is thus not surprising that they are unpopular with tropical fruit growers, especially as they may arrive in hundreds to feed on a ripening crop of fruit.

Microchiropterans

The second sub-order of bats is the Microchiroptera, usually called the insect-eating group. They are usually smaller than the members of the other suborder. They have poorly developed eyes, but an excellent echolocation system, which enables them to navigate and find food, even in pitch darkness. The bats that are most familiar to Europeans, Americans and Canadians are the vesper bats, which include the common little brown bats of the genus *Myotis*, the pipistrelles of the genus *Pipistrellus*, the yellow bats, and the hairy-tailed bats. Although most female bats give birth to a single baby at a time, several vesper bats have twins, and the female of the American red bat (*Lasiurus borealis*) is known to have triplets and even quads.

Several bat families are named after the peculiar structure of the bats' noses. Around the nostrils fleshy growths have evolved ranging from a simple spike to incredible structures. There are just too many of them to name all the groups but the common names will give an excellent idea of the bats' facial features. For example, there are tube-nosed bats, mastiff

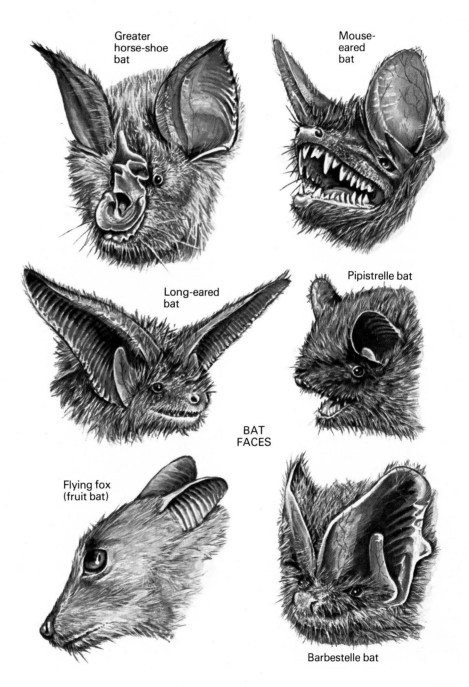

Greater horse-shoe bat

Mouse-eared bat

Long-eared bat

Pipistrelle bat

BAT FACES

Flying fox (fruit bat)

Barbestelle bat

Left: *Greater horseshoe bats* (Rhindophus ferrum-equinum) *at rest. Their name comes from the shape of the peculiar expansion of skin surrounding the nostrils.*

bats, spear-nosed bats, leaf-nosed bats, flower-faced bats, slit-faced bats and horseshoe bats. The majority of these bats feed on insects, catching them on the wing by scooping them up in the membranes of the tail which acts as a scoop. However, there are several other ways of feeding.

The most specialized method of feeding is shown by the vampire bat (*Desmodus rotundus*), which is confined to the tropical areas of Central and South America. It only eats blood, which it obtains in the true vampire fashion by biting large mammals or large birds, in a naked area such as behind an ear or in the neck region. This wound is so carefully inflicted on the sleeping victim that it is not disturbed. The blood is lapped up, not sucked, by a long, narrow tongue which is licked out. Coagulation of the blood is prevented by a special chemical in the bat's saliva. The danger of the vampire is not just that it deprives its prey of blood, but it is known to be able to transmit diseases, such as horse-fever and rabies.

Many of the bats in this group have become adapted to eating food other than insects. The fish-eating bats (*Noctilia*), of Central and South America, swoop to catch a fish in their elongated hind claws. They even fall into the water sometimes, but are able to take off without difficulty, their short fur being waterproof and quick-drying. The false vampire (*Vampyrus spectrum*) is not a true

Above: *The face of a vampire bat, showing the extremely sharp incisor teeth which are used to make the small cuts for obtaining the blood. The blood is lapped up by its long, narrow tongue flicking in and out.*

vampire, but does eat animals such as smaller bats, rodents and birds. The prey is seized by the free claws and taken back to a hollow tree roost to be devoured at leisure.

Echolocation by bats

The insectivorous group of bats have large ears and complicated skin-folds and growths on their noses which are vital for finding their way about in the pitch darkness of caves or on their nightly feeding trips. As they fly, they emit powerful bursts of very high frequency sound which is inaudible to the human ear. The bats listen for echoes which are reflected from objects in their flight path. This technique has evolved to such a fine degree that they can detect fine twigs or telegraph wires or a flying moth which they may capture for food. The system, which is similar to sonar, allows the bats to fly and hunt for food, even on the darkest nights.

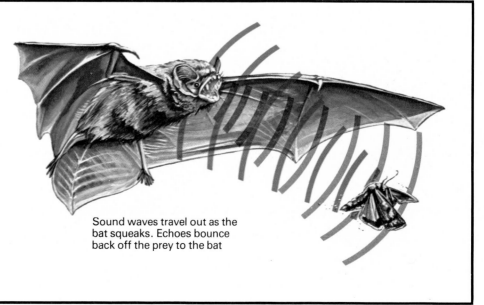

Sound waves travel out as the bat squeaks. Echoes bounce back off the prey to the bat

Primates

(ORDER PRIMATA)

The primates are in many ways the most interesting group of animals from man's point of view, because they form the order to which he belongs. There are about 200 species of primate and they are found mainly in the warmer parts of the world, although man (*Homo sapiens*) is nearly world-wide in distribution.

The living primates evolved from ancient primitive insectivores and so the living insectivores are the closest existing relatives of the primates. The trend in primates is towards a more active, diurnal, tree-dwelling life, with the sense of sight becoming more important than smell. The hands and feet are adapted for an arboreal life by being able to grasp, and the claws in many species have become flattened nails. The outstanding characteristic of primates is the great development of the brain (especially the cerebral hemis-

pheres which control the ability to reason), which reaches its peak in man. This gives a great deal of intelligence and reasoning power to the higher primates such as the chimpanzees, gorillas and other apes. The less advanced primates include the lemurs of Madagascar, the bushbabies of Africa, the lorises and tarsiers of southern Asia, and the tree shrews.

The 20 or so species of tree shrews of southern Asia and the islands of Sumatra, Java, Borneo and the Philippines are well named as they are arboreal and skilled at climbing. Externally they do not look like any other primate and, indeed, they show strong similarities to shrews. They have a squirrel-like tail and body, and they eat like squirrels, sitting on their hind legs and holding the food in the front paws. Tree shrews are omnivorous, eating a wide variety of food from seeds, leaves, and ripe fruit to worms and insects. The female gives birth to twins, usually, and it is interesting to note that she visits them in their nest, to feed them, only once every 48 hours.

Above: *Neither monkey nor insectivore, the tree shrew is regarded as a primitive primate. Squirrel-like, it is a skilled climber.*

They gorge themselves until they are so full that they are pot bellied, but this allows them to survive happily until the next feed.

Lemurs of Madagascar

The 16 species making up the family of lemurs are all found on Madagascar and the Comoro Islands. They usually inhabit wooded areas. Unfortunately these areas have been destroyed to such an extent in recent years that the majority of lemurs are now in danger of extinction. Lemurs vary in size from that of a mouse to that of a small dog. They have fox-like muzzles, large staring eyes, long heavily furred tails and slender bodies and limbs. The fur is usually soft, thick and

Right: *A ring-tailed lemur takes a rest in a tree on its island home of Madagascar. The striking tail is used to signal among the members of its social group, as well as a 'scarf' when it sleeps.*

woolly, the colour being plain or patterned depending on the species.

The ring-tailed lemur (*Lemur catta*) has its tail ringed with black and white, which is most distinctive and is used as a signalling device within a family group. This lemur is the one most frequently seen in zoos, and it breeds well in captivity. On its island home, social groups of 12 or more live in thinly wooded country where there are rocky outcrops. Ring-tailed lemurs spend much time on the ground, unlike most other lemurs which are largely arboreal.

A most attractive lemur is the ruffed lemur (*L. variegatus*), which has piebald, fluffy fur and a white ruff. Not all lemurs are as attractive as the ring-tailed and the ruff species. The brown lemur (*L. fulvus*) is a uniform brown colour apart from its black face, while the red-bellied lemur (*L. rubriventer*) is also brown but has light patches. The smallest lemur is the lesser mouse lemur (*Microcebus murinus*) and indeed this mouse-sized mammal is the smallest living primate. They are normally very active nocturnal animals, but in the dry season they roll up into a ball and sleep in a torpid state until the wet season

begins and food becomes plentiful again. This is known as aestivation and is analogous to the hibernation of animals living in cold climates. This is extraordin-

ary behaviour for a primate.

The sifakas and indrises are large, long-haired lemurs. They cling upright to the branches of their Madagascan trees, jumping

Flying lemurs

The flying lemurs (*Cynocephalus* species), also called colugos or gliding lemurs, are placed in an order of their own, the Dermoptera, which means 'skin-winged'. There are only two species, which are confined to southern Asia, Sumatra, Borneo and the Philippines. They do have affinities with the lemurs of the primate order but zoological classification experts insist that they should have their own order. These creatures do not fly but can only glide by means of the extensive folds of skin which embrace all the limbs and meet under the neck. The animals have lost their power of walking and, when stranded on the ground, they have to crawl in 'bat-fashion' to the nearest tree.

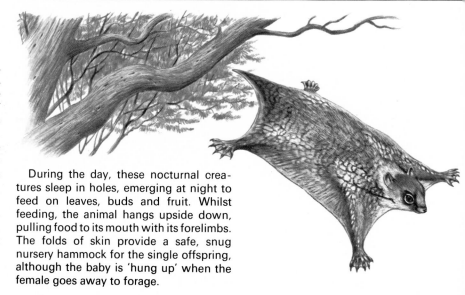

During the day, these nocturnal creatures sleep in holes, emerging at night to feed on leaves, buds and fruit. Whilst feeding, the animal hangs upside down, pulling food to its mouth with its forelimbs. The folds of skin provide a safe, snug nursery hammock for the single offspring, although the baby is 'hung up' when the female goes away to forage.

Left: *Demidoff's bushbaby. The huge eyes are for night vision, and large ears pick up the sounds of insects moving.*

from one to the next, and gripping with their padded hind feet which have the big toe opposable. The elongated hands have the short thumb slightly opposable, for gripping purposes. The indris (*Indri*), a large lemur, is held sacred by the local natives and so they never kill it, but even so it, as with all the primates of Madagascar, is quite rare and in danger of extinction since its habitat is being destroyed.

Pottos and lorises

The pottos (*Perodicticus potto*) and angwantibo (*Arctocebus calabarensis*) of Africa, and the lorises of southern Asia are the slow movers of the primates. All are tailless, but they have good gripping hands and feet which make them excellent climbers. They move hand over hand through the branches, often travelling upside down but rarely losing their balance because they always make sure that three limbs are grasping the branches. The angwantibo is the most agile member of this group, and has been seen to turn somersaults by walking backwards between its hind legs.

The bushbabies or galagos are the long-legged acrobats of the loris family. They are fast movers, covering spectacular distances with their jumps from tree to tree, 5 metres being a fairly easy leap. When they come down to the ground they hop like small kangaroos. They are nocturnal primates with large, forward-facing eyes and soft, thick fur. The long tail is an excellent balancing organ. The Senegal bushbaby (*Galago senegalensis*), the most familiar zoo bushbaby, lives in small family groups, spending much time grooming another bushbaby with its comb-like lower incisor teeth. The smallest

bushbaby is Demidoff's bushbaby (*G. demidovi*), which fits easily into a man's cupped hand.

Bushbabies mainly eat insects such as grasshoppers, holding one in the hand to bite off the head before devouring the body. A female bushbaby usually gives birth to 1 or 2 babies after about 120 days gestation. Although they do sometimes cling to the female's body when she moves about, usually they are left clinging to a branch until she returns.

The small, huge-eyed, large-eared tarsiers (*Tarsius* species) of the Philippines, Borneo, Sumatra and Celebes, are strange primates. They have a mixture of the characteristics of the lower primates and those of the highly advanced primates. The flattened face, round skull and the eye sockets link them to the true monkeys.

Below: *A tree-dwelling tarsier of south-east Asia has elongated fingers and toes tipped with rounded, soft pads that enable it to grip almost any surface. It springs on insects and grasps them with its hands.*

New World monkeys

Monkeys are usually divided into two groups, those that live in the New World, and the rest which inhabit the warmer parts of Africa, India and Asia. The New World monkeys or cebids have wide noses, the nostrils of which open towards the side rather than towards the front, in contrast to the narrow noses and forward-opening nostrils of the Old World monkeys and the higher apes. There are two groups of New World monkeys, the first including the spider, woolly, howler and uakari monkeys, and the other the small marmosets and ta-marins.

The titi monkey (*Callicebus*) and the related night or owl monkey, or douroucoulis (*Aotes tri-virgatus*), belong to the first group. The titi is active during the day, searching for insects and small animals. The night or owl monkey, as these names suggest, is active during the night. It has large, saucer-shaped, yellow eyes that enable it to find insects, other small animals, and fruit for food. During the day it curls up in a hole in a tree or among thick foliage.

Uakaris are rather strange looking monkeys. They are also called blushing monkeys because their red faces blush deeper when the animal is excited or upset.

They are exceptional in having short tails, which contrast with the very long tails of other New World monkeys. The 3 species are the blackheaded, red and bald uakaris, all with long shaggy gingerish hair. They live in small bands in the tops of trees, where they feed on fruit, nuts and leaves, and they rarely descend to the ground.

Howler monkeys (*Alouatta*) are found in the northern part of South America and in Central America. They are quite large with long prehensile tails, the end of which is naked and so excellent for grasping branches. These monkeys are so named because they have a sac or sound-box in their throats which amplifies their voices. The resonating voices, often issued as a chorus, can be heard 2 to 3 kilometres away. The howler monkeys are territorial, each group having its own tree area and they move through the tree tops along regular routes.

An attractive and charming monkey of the tropical forests of South America is the squirrel monkey (*Saimiri sciureus*). It moves about in small family parties or in larger groups of up to 100 monkeys. They are very active, gregarious animals, wandering through definite routes of vegetation near rivers, searching for food which ranges from flowers and fruit to insects and frogs. They will even eat a small bird if they succeed in grasping one.

The spider monkeys (*Ateles* species) are well named, on account of their slender bodies, long, slim limbs and their spider-like gait as they move along branches or over the ground. The extremely long prehensile tail is used as an extra limb and sometimes the monkey will hang suspended from its tail

Above: *A spider monkey shows off its prehensile tail to perfection.*
Right: *A tiny cotton-top tamarin uses its headplumes as part of its signalling and recognition behaviour.*

using its hands and feet together to hold fruit, nuts and leaves. The woolly monkeys (*Lagothrix* species) are related to spider monkeys but have much denser fur and a rather 'crew-cut' appearance. The habits of the two types are similar.

The tiny, twittering marmosets and tamarins are all found in the Amazonian region. The 33 or more species are all dwarf monkeys with shrill voices and a quick, scurrying gait. Their hands and feet are long and nar-row and end in claws, not nails as they do in most monkeys. They are squirrel-like in behaviour and not only move like these rodents but often lie on their bellies on a branch with their limbs hanging down either side. All of the marmosets and tamarins have long, hairy, non-prehensile tails. Many of them have weird head adornments. For example, the common marmoset (*Callithrix jacchus*) has white ear tufts, the black pencilled marmoset (*C. pencillata*) has black ear-plumes and the cotton top tamarin (*Leontocebus oedipus*) has head plumes. These features are important within the social group for recognition and signalling.

The most vividly coloured mammal

The golden lion marmoset (*Leontideus rosalia*) is the most brightly coloured of all living mammals with its shimmering coat of gold. The first living specimen seen in Europe was apparently owned by Madame de Pompadour. Why it is this colour is not known, but because it has been so much admired and therefore captured by man, numbers have fallen below 500. Also much of its forest habitat, in the coastal region of south-eastern Brazil, has gone to make way for agricultural plantations and housing projects.

An unusual aspect of most marmosets and tamarins, including the golden lion marmoset, is that after a long gestation period of over 140 days, twins are usually born and the father takes care of them, except at feeding times when they are transferred back to mother. He wears them scarf-style round his neck when they are tiny but later they cling to his back.

Old world monkeys

There are over 60 species of Old World monkeys. Most of them have a tail but not one can use it to grasp with. They appear to be higher in the evolutionary tree than their New World relatives and probably have some degree of colour vision. Most are arboreal in habit, but the baboons and macaques have abandoned a tree-dwelling life to roam around in either a grassy or rocky habitat. Like man, all have 32 teeth. In some monkeys the thumbs are very small or absent altogether, but in those monkeys where they are fully developed they are always opposable to the finger, giving good grasping hands. The Old world monkeys are all placed in the family Cercopithicidae and due to the enormous variety within the family they are divided into two groups.

Macaques, mangabeys and baboons

The first group contains the macaques, mangabeys, baboons, mandrills, and guenons, all of which have long muzzles and stomachs that are of fairly simple structure. Macaques are cat-sized with robust bodies and short legs. All have cheek pouches in which they temporarily store food. The rhesus macaque (*Macaca mulata*) is the best known, as it is favoured for medical research and space exploration. It is a fairly large monkey with a shortish tail and yellow-brown fur. In the wild, it is a social animal, moving around in large noisy troops throughout its north Indian and south-east Asian range. The only European monkey is the Barbary ape (*M. sylvana*), which inhabits the mountains of Morocco and Algeria and the rocky slopes of Gibraltar, where it is well-fed on titbits from tourists.

Mangabeys live in the tropical forests of Africa and are long-muzzled, dark-coloured monkey with large cheek pouches and slender bodies and long tails There are some 4 species and al make great use of grimaces ir communicating with one an other. They also signal with thei white eyelids. The most distinc tive is the white-collared or red capped mangabey (*Cercocebus tor qutus*), which is thought to be a race of the sooty mangabey.

Baboons are large, dog-like mon keys that have adapted to a life

Rhesus macaque monkeys are widely used in behavioural experiments. This baby monkey is being tested to discover the results of maternal deprivation. Parted from its mother, it is offered two mother substitutes: one is warm, soft and comforting, the other is cold, made of wire but provides food. It is interesting to discover that the baby first goes to the cloth-covered model for comfort, and only later, when it feels secure, does it move to the other model for food.

on the ground. The 6 species all have enlarged and elongated muzzles which house huge teeth. The yellow, chacma, olive and guinea baboons are all very much alike in appearance, differing only in size and colour. The hamadryas or sacred baboon (*Papio hamadryas*) is very distinctive with silvery-brown hair and a magnificent cape over the shoulders of mature males. The gelada baboon resembles it, as it too has a cape of long hair.

The mandrill (*Mandrillus sphinx*) and the drill (*M. leucophaeus*) are related to the baboons and like them have long dog-like muzzles with huge teeth. The mandrill is one of the strangest looking primates. Its exposed parts of skin are the most highly coloured of any living mammal. The long nose is a vivid red, while the strikingly ridged skin of the cheeks is an intense bright blue. The fur of the head and body is mostly brown although mature males have yellowish white beards and bellies. However, the hind quarters of a mandrill are also naked and vivid, with red and blue markings which in places blend to give a delicate shade of mauve. Drills are very similar to mandrills in shape, but are smaller and much less brightly coloured, their naked faces and rumps being jet black. They both live in the forests of central West Africa where they spend most of their time on the forest floor. They live in family groups and feed on a wide variety of foods.

Guenon is the name given to some 10 species of monkey that belong to the genus *Cercopithecus*. The grass or green guenon (*C. aethiops*), which is also called the vervet or grivet is typical of them. It is the most common of all African primates and is a graceful, cat-sized monkey with a slender body and long limbs and tail. These busy monkeys prefer the savanna lands and anyone fortunate enough to go on an African safari is sure to see many of them moving around, often in large troops. The smallest guenon and indeed the smallest monkey of the Old World is the attractive talapoin monkey. It is half the size of most guenons and is only about 32 centimetres long. Like the grass monkey it is greenish yellow in colour and has white rings around its eyes and a white chest.

Below: *Male splendour of the mandrill. The red nose and brilliant blue, ridged cheeks help to attract females to him.*

Langurs of tropical Asia

The 18 or so species of langur form the second group of the family Cercopithecidae. They all live in the tropical forests of southern Asia. Although they are in general uniformly coloured, the upper part of the head often has coloured adornments. The douc langur, for example, has a brown head, with a bright chestnut band below the ears, a bright yellow face and white whiskers. Other langurs have coloured crowns or crests.

The young of the spectacled langur has a brilliant golden-orange coat, which primatologists believe helps the parents and family group to keep the animal within sight during its infancy. The coat gradually

Right: *Spectacled langur* (Presbytis obscura) *of south-east Asia surveys the scene. Any offspring are born a brilliant orange in colour which probably serves to keep the parents informed of its whereabouts.*

Below: *Colobus monkeys relax in a social group in their African forest home.*

changes to the adult's dark colour with increasing age.

The true oddity of the langur group and indeed the whole primate group is the weird looking proboscis monkey (*Nasalis larvatus*). The mature male has a swollen bulbous nose that is long and pendulous. It is less well developed in females, and in youngsters it is even turned upwards until the animal begins to mature. These strange looking primates live in family groups in the low mangrove swamps of Borneo, and, as with all langurs, their diet consists mainly of leaves.

The colobus monkeys (*Colobus* species) are long-haired, slender-bodied animals of the forests of tropical Africa. They are related to the langurs and are also leaf-eating in diet. Three species are recognized, the red, the green, and the black and white. The fur of the black and white colobus is handsomely marked and is very long and silky. This has led to thousands being slaughtered, especially at the end of the nineteenth century, for the fashion trade with coats, dresses and shawls being made from the striking skins.

Apes

The ape family contains 10 species, which are found in the tropical forests of Africa and southeast Asia and nearby islands. They are the closest living relatives of man. The term 'great apes' is usually given to orang-utans, chimpanzees and gorillas, which are all large man-like, or anthropoid, apes with protruding jaws. The other apes are the agile gibbons and siamangs. All apes are tailless and have very long arms.

Right: A Sunda Island gibbon shows its grasping hands and feet, that enable this highly agile small ape to move through the branches at great speed. It swings by hooking its hands over branches.

They have highly developed brains and show a great deal of intelligence, more so than any other living animal apart from man.

The 6 species of gibbon live in the rainforests of Sumatra and southern Asia. The gibbons are all similar in shape and size and have very slender bodies and long, slim limbs. The arms are longer than the legs and are used to swing through the trees from branch to branch. They do not grasp the branches as they move at high speed but simply use the long slim hands as hooks. Over 3 metres can be covered in a single swing. This method of movement is known as brachiation. Sometimes they will run along a branch on their short legs, using their arms to balance. They also use this method when travelling along the ground.

Gibbons live in small family groups, consisting of a breeding pair and their offspring. They defend their territory from other gibbons and intruders, these being warned off by loud and echoing 'hoo-hoo-hoo' calls. The family searches for food of fruit, leaves and buds, with an occasional insect, young bird or egg, during the day. At night they sleep erect in trees, huddling together for comfort.

The siamang (*Hylobates syndactylus*) is the largest member of the gibbon group. It is also the most vocal, due to a huge, naked, reddish-brown vocal sac. This is inflated prior to calling and tremendous whooping cries are produced.

The orang-utan (*Pongo pygmaeus*) lives in the low-lying forests of Borneo and Sumatra where it is in danger of extinction because its forest home is being destroyed at a great rate. Also there is a great deal of poaching of youngsters, the mothers usually being shot. This is illegal as the orang is officially protected to try and prevent its extinction.

Orangs are easily recognized by their long, red, shaggy coats. In the male, the almost naked face develops cheek pouches and a throat pouch. It is no wonder that it is known by the natives as 'Old-Man-of-the-Woods'. An adult male stands 1.5 metres high and can weigh up to 100 kilograms, while the female is smaller, rarely weighing more than 50 kilograms. The orang's arms are much longer than the legs and are used to brachiate slowly through the trees. Its hands and feet are long and narrow to give a good hold.

Orang babies are most attractive and many zoos exhibit these charming youngsters. Today, zoo specimens have usually been bred in captivity, but occasionally orphan babies are sent to zoos when they have been confiscated from poachers and illegal traders. Young orangs are born after 8.5 months gestation and the single baby is suckled for 18 months.

The chimpanzee is perhaps the favourite ape because the young chimp is so human-looking, so mischievous and playful. The single species of chimpanzee (*Pan troglodytes*) lives in the forests of central West Africa in small family groups. A small very dark race, the bonobo, is found in the south Congo regions and some regard it as a distinct species (*P. paniscus*).

The loveable chimp

The chimp is lighter in weight than the orang-utan, with a smaller head, shorter legs and arms, and a naked face, which is human-flesh-coloured in youngsters but darkens with age. Large flesh-coloured ears protrude and it has a coat of dark brown to black fur. Although the chimp does climb trees and usually sleeps in a crude nest that it constructs with entwined branches and leaves, it spends its life, for the most part, on the ground. A small family party roams the forest in search of ripe fruit and leaves, insects, small bird nestlings and eggs. They walk on all fours, the soles of the feet and the knuckles of the hands in contact with the ground. Sometimes chimps will stand upright or walk, holding their hands to the side or even above the head for better balance.

The social life of the chimp is well developed. They communicate by touching and use a wide range of vocal sounds to express their feelings. They are very intelligent animals and research into their behaviour in captivity and the wild has shown that they can learn to do certain tasks and also resolve problems using logical reasoning.

The great but gentle gorilla

The gorilla (*Gorilla gorilla*) is the largest of the primates and in the past has given rise to an enormous number of legends. Early explorers to its native home of the forests of central West Africa reported huge hairy 'men' living there. The gorilla was not properly known to science until 1847 but today quite a lot is known about its behaviour in the wild since research students have studied it in its rather inaccessible jungle habitat.

Despite stories of the gorilla's ferocity, it is relatively shy, gentle and retiring. It becomes aggressive only when irritated or molested, and then does indeed bang on its chest in true 'King-Kong' style to give warning of its attacking intentions to the enemy.

An average adult male gorilla stands at just under 2 metres and can weigh up to 225 kilograms. A mature male has a large bony crest on the top of his skull, giving him the appearance of wearing a furry helmet. The chest is very broad and the head seems to emerge straight from the chest because the neck is short. An elderly male usually goes silvery grey on his shoulders and back and is called a 'silverback'. Females are shorter and less heavy than the males. A family group, led by a dominant old 'silverback', spends most of the day trekking through the thick undergrowth of the jungle, stopping when it finds succulent leaves or ripe food. Although they have great long canines, gorillas are strict vegetarians. At night the family climbs the trees and makes crude nests out of twisted branches and leaves. The young sleep in the highest position as they are not only lighter than the others but this is the place safest from the leopard, the gorilla's only enemy. The only member of the family to sleep on the ground is the mature leader who is usually too heavy to climb. He usually sits with arms folded at the base of the 'sleeping tree', dozing but ever alert for intruders.

A single young is born after 8 to 9 months, and is nursed in the mother's arms in a similar way to a human baby. It receives much care and attention from the family and it is walking by about 10 months. A female is sexually mature at about 7 to 8 years, a male taking about a year longer. Gorillas have lived in captivity up to the age of 34 years and it is thought by some that the life span in the wild could be as much as 50 years.

Right: *Silverback leader makes sure all is clear for a mother and youngster to cross.*

Above: *A silverback gorilla is a mature old leader that usually dominates his* family, *guarding it against intruders and seeing off any rivals to his group.*

Above: *Enjoying a feast of tender leaves and shoots, a large family of gorillas are* still quite alert. *Although equipped with large canines, they are strict vegetarians.*

Whales

(ORDER CETACEA)

Whales, dolphins and porpoises belong to the order Cetacea, a group of mammals which abandoned a life on land for a totally aquatic existence some tens of millions of years ago. They still have the mammalian characteristics of being warm-blooded and suckling their young, although their shape is more fish-like than mammal-like. Whales have lost the thick fur coat that most mammals have to keep in the body heat. Most young whales, however, do have a few bristles round their mouths. To retain body heat, the 90 or more species have a very thick layer of a very oily substance, called blubber, under the skin. The blubber also helps to streamline these mammals so that they are the most suitable shape for moving through the water with least resistance.

As whales, dolphins and porpoises never leave the water, they are more perfectly streamlined and adapted for aquatic life than are the seals and their relatives. The forelimbs of cetaceans have become shortened and broadened so as to become 'flippers'. The hind limbs have disappeared completely and only vestiges of the hip bones remain. The most noticeable feature that distinguishes whales from fishes is that the great tail flukes are placed horizontally, whereas the tail-fin of the fish is upright. So in swimming, the cetacean tail moves up and down rather than from side to side as in the fishes. Whales breathe by lungs, as do all mammals, and although they are able to spend a considerable time under the water, they have to surface eventually to get a new supply of air and rid themselves of the used air. The action of puffing out the breath from the nostrils

Above: *A killer whale surfaces to take a breath. This toothed whale kills seals, penguins and porpoises, ripping them apart.*

that open at the top of the whale's head is called spouting. As the warm breath meets the surrounding air it condenses into a sort of mist and looks like a fountain of water. Each whale species has its own type of spout and can be identified from a distance just by the spout's shape.

Toothed whales

Whales, which are found in all the oceans, are divided into two main groups. Firstly, there are the toothed whales, the odontocetes, which, as their name suggests, have teeth. These teeth are often very numerous, usually conical in shape and simple in structure. The group includes the sperm whale, narwhal, killer whale, dolphins and porpoises.

In the sperm whale (*Physeter*

Cetacean babies

Cetaceans usually give birth to a single offspring after 11 to 16 months gestation, many whales migrating to warm tropical waters to give birth. The newborn young is usually a quarter to one third the length of the mother. Immediately after being born in the water, the baby must reach the surface to obtain a supply of air. The female dolphin pushes her baby to the surface, and other dolphins, 'aunts', are usually close by to help. The mother floats on her side so that the baby can suckle and breathe at the same time. Her milk is rich and the infant grows rapidly.

catodon) the males sometimes reach 20 metres in length. When attacked the sperm whale can be very dangerous and in the old days, when harpoons were thrown by hand from small boats, the men were often thrown into the sea as the boat was upset by the whale's thrashing. Melville's Moby Dick, which bit off a man's leg, was a sperm whale.

The ferocious black and white killer whale (*Orcinus orca*) feeds on seals, porpoises, penguins and other seabirds, as well as on large quantities of fish. It is without doubt a fearsome predator of the oceans. One killer caught in the Bering Sea was found to have 32 adult seals in its stomach. They often hunt in packs of up to 40, and have been known to attack the giant blue whale, tearing at the victim until it bleeds to death and can be devoured in safety.

A number of small toothed whales, especially those with beak-like jaws, are called dolphins or porpoises. They are often seen in schools, accompanying ships for kilometres, playing gracefully around the bows. The bottle-nosed dolphin (*Tursiops truncatus*) has become famous in recent years as the star of the oceanarium exhibitions. These animals display a remarkable degree of intelligence and a constant need for activity and novelty.

Whalebone whales

The second group, the whalebone whales or mysticetes, have, instead of teeth, a row of long, horny plates, called the 'baleen' or whalebone. The baleen hangs down from the roof of the mouth on each side of the huge tongue. These whales, which include the giant blue whale, the largest animal alive today, feed by taking in a huge mouthful of water together with any tiny marine animals that may be in it. They force the water out again through the baleen which acts as a sieve: the small animals are held back and these are swallowed. Old whalers found some of these whales easier to catch than most because they swam slowly and floated when dead. They were also very profitable with their long baleen plates, and so the men named them the 'right whales' as they were the right whales to catch.

The blue whale (*Balaenoptera musculus*) is a member of the rorqual whales or fin whales. It is the largest of all mammals and reaches a length of nearly 30 metres and can weigh over 100 tonnes. Other rorquals include the humpback and sei whales.

Below: *Star of most oceanariums, a bottle-nosed dolphin performs.*
Bottom: *Wild and free, bottle-nosed dolphins jump around the photographer's boat. They find their way by echolocation.*

Above: *A giant anteater of the South American pampas strides along in search of a termite mound. It will rip this open with sharp, strong front claws, that are curled under as it moves, giving it a curious pigeon-toed gait. A sticky tongue shoots out of the long snout to trap the numerous insects.*
Left: *The upside-down world of a two-toed sloth. Its hair grows up the body and arms so it hangs down to allow tropical rain to run off.*

Edentates – the toothless ones

(ORDER EDENTATA)

Armadillos, sloths and anteaters belong to the order Edentata, which name means 'without teeth'. This is rather inaccurate in that only the anteaters lack teeth and the giant armadillo actually has more teeth than any other mammal, except some of the whales. At first, the three types of animal look rather different from each other, but extinct fossil forms have shown they did have common ancestors in the past. These evolved in South America, where almost all the surviving members remain.

Anteaters

The anteaters of the family Myrmecophagidae are the most specialized insect-eaters among the mammals. The 4 species all inhabit central and northern South America, and they are built on a similar plan. The front part of the skull is drawn out into a tube with a mouth which opens at the end and is so small that it only allows room for the 50 centimetre long, sticky, worm-like tongue to be flicked in and out. The tongue laps up termites and ants. The front limbs have very powerful digging claws that are sometimes used in defence, although they are usually used to tear down termite colonies.

The giant anteater (*Myrmecophaga tridactyla*) prefers open forests and grasslands and is quite remarkable in appearance. It is up to 2 metres long from the tip of its tapering, tubular snout to the end of the long bushy tail. The grey-brown body has a bold black and white wedge-shaped pattern on the throat and shoulders, which breaks up the body outline and thus helps to hide the animal within its surroundings. This anteater walks on its knuckles so that the strong claws of the forelimbs are kept sharp for digging. A baby anteater clings to the rump and tail of a parent, riding jockey style.

The smaller tamandua or lesser anteater (*Tamandua tetradactyla*) prefers more thickly forested areas and spends much of its time in trees. The long prehensile tail helps it in its arboreal life. It feeds at night on ants and termites, often attacking colonies in trees. The dwarf anteater (*Cyclopes didactylus*) is slightly larger than a squirrel, and is exclusively tree-dwelling, living high up in the thickest forests.

Sloths and armadillos

Some 7 species of sloth make up the family Bradypodidae. They all live in trees, nearly always climbing and hanging in an upside down position. They use the long, curved claws on their limbs as hooks. The different species fall into two groups, which are identified by the number of claws and digits on the front feet. The two-toed sloths (*Choloepus*) and the three-toed sloths (*Bradypus*) look alike and it takes a close examination to identify them.

Three-toed sloths are very fussy feeders, preferring the leaves and buds of the cecropia tree, which is related to the mulberry and hog plum trees. Two-toed sloths, on the other hand, will feed on a variety of vegetation. The single young clings to its mother's chest as she moves about.

The 20 or so species of armadillo are grouped in the family Dasypodidae. The largest armadillo is the giant armadillo (*Prio-dontes*) which measures just under 2 metres in length and can weigh as much as 50 kilograms. The smallest is the fairy armadillo (*Chlamyphorus*) which is about 1.3 centimetres long and spends most of its time underground, its eyes and ears being very similar to those of moles.

Armadillos range all over South America except for the inhospitable cold south and the high Andes. They are armoured with scaly, bony plates, yet are able to move as these plates are connected by tough and flexible skin. All have strong claws which help them to dig for insects and other small animals on which they feed, to make burrows, and to escape from predators. Their armour provides the other defence against predators, and some species, such as the three-banded armadillo (*Dasypus*), can roll themselves into a tight ball, all the soft parts being protected.

Below: *A nine-banded armadillo* (Dasypus novemcinctus) *has plates connected by skin to allow free movement. It will burrow quickly to escape predators or roll up into a defensive ball, protecting its soft underside.*

Carnivores – the meat eaters

(ORDER CARNIVORA)

All mammals, and indeed any animal that feeds on another live animal, are in effect carnivorous (for example insectivores and insect-eating bats) but in the zoological sense the term carnivore refers to members of the order Carnivora. It includes the cat, hyena, civet, dog, raccoon and bear families. Close relatives are the seals and sea-lions which were once classified as Carnivora, but are now placed in their own order. The carnivores are divided into two main groups, the cats, hyenas, genets, civets and mongooses in one, and the dogs, foxes, wolves, bears, raccoons, pandas, weasels and badgers in the other.

Cats

The cats of the family Felidae usually have slender bodies, a long tail and strongly pointed and curved claws. The claws can be retracted inside protective sheaths, so that they do not rest on the ground while the animal is moving. In this way they are kept sharp for use as weapons of defence or attack, or as aids to climbing trees. The worn claws are re-sharpened by rubbing off the outermost horny layers by clawing the bark of trees. The heads of cats are rounded with a short muzzle. They all have an excellent sense of hearing and good binocular vision, important in finding and capturing their prey. There are about 36 species

Above, left: An impressive portrait of a lion with his mane identifing his sex.
Left: A lioness allows her cubs to attempt to feed on a freshly killed zebra. These cubs still have spotted coats which help to camouflage them in long grass and undergrowth where they hide for hours.

of cat found living almost all over the world, except Australia and the Antarctic.

It is usual to refer to the lion, tiger, jaguar, leopard and snow leopard as the 'big cats'. The lion (*Felis leo*) used to be found throughout much of the Old World, but today it is found only in Africa, south of the Sahara, with a small number living in the Gir Forest in north-west India. Lions are the only truly social cats and normally live in prides consisting of one or more adult males, a large number of females and their offspring. A male is identified by the large long-haired mane on his head, neck and shoulders, which sometimes extends along the belly region. He is up to 3 metres long from nose to tail-tip. The female is maneless and smaller than the male. The cubs are usually spotted, a camouflage adaptation for their first months of growth. Females tend to retain spots on their legs and bellies.

Lions hunt mainly antelopes and zebras, but occasionally will attack weak or young giraffes and buffalo. They never prey on ele-phants, hippos or rhinos, however. They hunt silently, usually in groups, and often the females do most of the work. They lie in wait, often at a waterhole, and then when herds come to drink, with a quick sprint and a leap, they attack without warning. When not hunting, the members of a pride spend most of their time dozing. Over 18 hours a day is commonly spent asleep.

The tiger (*F. tigris*) roams the forests of Asia, Sumatra and Java, but is in a rapid state of decline in most regions and may not survive this century, in the wild. In general, tigers are very similar in build to lions, although certain races such as the Siberian tiger are much larger. Tigers have rust-yellow fur with black transverse stripes, except on the belly and chin where it is white.

Tigers usually live and hunt alone, coming together in pairs for mating and then going their separate ways again. Recent research has shown that tigers do meet up with one another more often than was previously thought, but this is usually on the borders of a tiger's established territory. They hunt silently at night, stalking the prey such as antelope, pig or deer and then bounding the last few yards. Two to 4 cubs are born after about 108 days gestation, and are then reared and cared for solely by the mother. She teaches them the art of hunting and killing prey.

Leopards and jaguars look very similar to one another. The jaguar (*F. onca*) inhabits the forests of the southern United States, and Central and South America, while the leopard (*F. pardus*) ranges over much of Africa and Asia. The jaguar is slightly larger than the leopard, but both are smaller than lions and tigers. Although both leopard and jaguar are spotted, on closer examination it can be seen that the arrangement of the spots varies. A jaguar's rosette (a ring of spots) usually has another spot inside, while a leopard's does not. Another distinguishing point is that a jaguar's tail is shorter than a leopard's.

Below: *A snarling tiger shows its huge sharp canines that will rip at the neck of a deer or antelope, cutting blood vessels.*

Left: The ocelot of the forests of Central and South America is an agile climber and hunter, taking small mammals and birds it surprises with speedy jumps. It is now a protected species.

Leopards eat mainly small antelopes, jackals, monkeys and birds, usually eating the entrails and limbs first. Often they drag their prey up into a tree 'larder' so that other carnivores will not steal it. The jaguar hunts South American rodents, such as agoutis, and peccaries, deer and often domestic animals.

The snow leopard or ounce (*F. uncia*) is an extremely rare cat living in the mountains of Altai and the Himalayas, in central Asia. It is strictly protected these days, but numbers are very low because in the past it has been relentlessly pursued for the long, thick, beautiful fur that was used in women's fashions.

Pumas and small wild cats

The powerful puma (*Felis concolor*), also named the cougar or mountain lion, is slightly smaller than a leopard and inhabits mountains, plains, forests and deserts of America, from Canada to Patagonia. Numbers have been reduced by man's relentless hunting and killing, and in most regions it has retreated to areas relatively uninhabited by man.

The small wild cats include the northern lynx (*F. lynx*) of Eurasian and North American forests, the bobcat (*F. rufus*) of North America, the ocelot (*F. pardalis*) of the forests of Central and South America and the serval (*F. serval*) of Africa. The lynxes have short tails, long legs and hairy tufts at the tips of the ears.

There are several species of small wild cat which have plain-coloured bodies but extremely elaborately patterned faces. These include Temminck's golden cat (*F. temmincki*) of southeast Asia and the African golden cat (*F. aurata*). They prey on

The fastest land mammal

The fastest four-legged land mammal is the cheetah (*Acinonyx jubatus*). It is the oddest of all cats and is placed in a separate genus of its own. Some zoologists even put it in a family of its own.

A cheetah has permanently extended claws on the paws, situated at the end of its long legs. It stalks its prey, which is usually a gazelle, and when close enough sprints at about 100 kilometres per hour in order to bring the prey down. However, if the agile ungulate can out-manoeuvre the cheetah over some 400 metres, it will escape, since the cheetah has to pull up out of sheer exhaustion. It cannot sustain such speed for long.

Until about 10 weeks old young cheetahs have attractive silver manes down their backs. There are usually 2 to 5 in a litter.

Above: *A Eurasian wildcat snarls in its snowy lair. Looking rather like a large, domestic tabby, this feline survives in remote mountainous areas of Europe, feeding on birds and small mammals.*

small mammals such as rodents and small deer, as well as on birds.

The ferocious European wild cat (*F. sylvestris*) looks very much like a large domestic tabby, but it has a very thick, hairy tail with a round tip and long white whiskers. The domestic cats probably originated from the caffre cat, a sacred animal of the Egyptians, some time before 1600 BC. Today, numerous breeds have been established.

Hyenas

Hyenas of the family Hyaenidae have longer front legs than hind legs, so that the back slopes towards the tail, giving them a skulking gait. They are more dog-like in appearance than cat-like but their internal structure relates them to cats. There are 3 species of the genus *Hyaena*, the striped, brown and spotted hyenas. They live in the open coun-

Right: *Scavengers by day but active, hunting killers during the hours of darkness, hyenas are watched closely by a vulture and marabou stork waiting for any left-overs of the young wildebeest.*

try of Africa, western Asia and southern Asia.

It used to be thought that the hyenas were the scavengers of the savanna, feeding on other carnivore left-overs. It is now known, however, through research studies in the wild, that hyenas are active hunters. At night, in large packs, they track down prey such as antelope or zebra, and ruthlessly kill the victim.

The aardwolf (*Proteles cristatus*) is a close relative which eats only termites and insects. It is an ungainly animal looking rather like a small striped hyena but with a long mane down its back and a bushy tail.

Below: *A North American bobcat strikes lucky, flipping a salmon from shallow waters. It usually lives alone, pairing briefly for mating, and is an excellent tree climber, swimmer and fighter, using teeth and claws as effective weapons.*

Left: *An African genet using its slicing cheek teeth to devour its prey.*
Below: *An Indian mongoose demonstrat[ing] that it can outmanoeuvre a snake until it can grasp it around the head and kill it.*

Civets, genets and mongooses

The family Viverridae is a large family of about 80 species and includes the civets, genets and mongooses. Most of them have small heads with pointed snouts, long bodies and long tails. Scent glands, found near the rump, are very important for marking ter-

ritory and communicating wit[h] other individuals, as well as fo[r] attracting partners in the breed-ing season.

Eight species of genet are foun[d] only in Africa, while the re[-] maining species, the common o[r] small-spotted genet (*Genetta genetta*), is found in Palestine an[d] southern Europe. The elongate[d] head is fox-like, the body i[s] weasel-like, and the legs, fee[t,] claws and tail are like those of [a] cat. They are nocturnal hunter[s] killing ground-roosting bird[s] such as guinea-fowl, and sma[ll] rodents. Despite the fact tha[t] their spotted coats camouflag[e] them well, most of the day is spen[t] hidden in a tree hollow.

Civets are similar to genets bu[t] have longer legs and spend mor[e] time on the ground. The Africa[n] civet (*Viverra civetta*) is quit[e] large, reaching 1.3 metres i[n]

length, and looks rather like a huge tabby cat with a pointed face. It raises the black hair on its back into a crest when alarmed. It often uses a deserted burrow, such as that of an aardvark, for a daytime sleeping hole, hunting during the night. Other civets are found in Africa and Asia. Palm civets are smaller than true civets and are omnivorous in diet. Most species are found in Asia.

The tree-dwelling binturong (*Arctictis binturong*) is civet-shaped but has very long, shaggy, black hair which sticks up from the ears in tufts. It is the largest and perhaps the strangest member of the civet family and eats mainly fruits and plants, probably because it is not as quick as its truly carnivorous relatives.

The 39 species of mongoose are found in the warmer regions of the Old World. They are the 'weasels' of their range, hunting small mammals, insects, young birds especially waterfowl, fishes and insects. Many have a great liking for birds' eggs: they will pick up an egg in the forepaws, dash it to the ground to break the shell, and then lap up the contents with relish. All mongooses are very agile and many can climb trees. The most famous species is the Indian mongoose (*Herpestes edwardii*), which can grapple with, kill and even eat the cobra.

Spotted cats and the fur trade

The well-loved and well-known spotted and striped cats are, on the whole, all in danger of extinction, mainly due to the fact that a lot of people seem to feel richer and more beautiful if they parade before others in the skins of the dead cats. Expensive fur coats on sale in most modern cities have led to the death of thousands of wild cats including the tiger, jaguar, oce-lot, cheetah, leopard, snow leopard and clouded leopard. Legislation by governments of many countries against the trade in most spotted cat skins and products has certainly given new hope to these graceful and exciting animals, but poaching and illegal trading will still go on, as long as there is a demand for fur products. In South America, for example, the Brazilian Government dictates that the jaguar is completely protected and may not be hunted for its skin, and trading in the spotted skins is also illegal. However, poachers are not deterred and cat skins can still be bought along certain stretches of the Amazon river.

It is thought that there may be no wild cheetahs in Africa at all by the end of this century. They need quite a large amount of living space on their savanna homes, and establishing several protected areas in Africa is a very costly business. Even in the national parks, the cheetah has to contend not only with poachers but also with other carnivores. In the Serengeti, for example, there are only about 150 cheetahs and these have to compete with some 600 leopards, 2,000 lions and 3,000 hyenas.

Weasel family

The other main stem of the large order Carnivora contains the weasel family, the raccoon family, the dog family and the bear family. The weasels, stoats and martens of the family Mustelidae are similar in some respects to the ancestral carnivores. There are about 68 species alive today, distributed across both the Old and the New World. Mustelids, as the family members are called, are in general smallish, long-tailed, short-legged creatures with their long bodies slung low to the ground. The term 'mustelid' comes from the scientific name for the one genus, *Mustela*, which

Below: A polecat is a larger, heavier, slower-moving animal than the related mink. A female will have 2 litters a year rearing her 5 to 8 babies alone. The domestic form of the polecat is the ferret.

includes the weasel and its close relatives. However, we also find such animals as skunks, badgers and otters in this family.

Weasels

There seems to be an endless list of weasels and the term is usually meant to include close relatives such as the polecats, minks, martens, tayras, grisons, striped weasels and the zorillas. All are excellent hunters and live entirely on the meat thus obtained, although one or two will take carrion. True weasels are very alike in form and colouration, apart from the stoat of Europe, which in winter in the northern part of its range turns pure white, except for its black tail tip, and is then known as the ermine. Although weasels are not very large, ranging between 15 centimetres and 60 centimetres in length, they are

enormously strong and fast-moving little beasts that can kill much larger animals. One weasel species is trained by tribesmen of northern Burma to kill large wild geese and even small goats. All weasels kill in the same manner, by biting through the main arteries in the neck of the unfortunate victim. One weasel will kill all the hens in a poultry house if it is not promptly caught.

Polecats are similar to true weasels but are larger, heavier and slower-moving animals, inhabiting the more open woodlands of Europe, Asia, North Africa, and North America. The European polecat (*Mustela putorius*) is almost certainly the

Below: A mink peeps out from its refuge in a hollow log. Its fur is modified for its semi-aquatic life, being oily and repelling water. Its hind feet are slightly webbed to help it swim.

animal from which the domestic ferret was bred. The albino ferrets are specially bred so that their white coats show up clearly when they are used in hunting for rabbits or rats. Like other members of the weasel family, polecats have large anal glands that are used to mark out the home range with the personal scent of the owner. The den usually smells strongly of its owner, which in England has led to the polecat being called the foul marten.

Minks are slightly shorter and more compact in form, and are famous for their thick, lustrous fur. It is the American mink (*M. vison*) that is farmed extensively and has been introduced into other parts of the world for this reason. In the wild it is usually found close to water and its hind feet are slightly webbed for better swimming. It eats mainly frogs,

fishes and crayfish, usually hunting at twilight or during the night. A female has one litter a year and the babies follow her for several weeks.

The martens are the best tree climbers, although the most famous, the Russian sable (*Martes zibellina*) spends most of its time on the ground. Martens, with their supple bodies and bushy tails, acting as balancing organs, are capable of catching arboreal prey such as squirrels, and ground living animals such as rabbits, partridges and chipmunks.

The glutton is the largest weasel and certainly deserves its common name because it gorges itself with prey such as caribou and deer. Even its Latin name, *Gulo gulo*, sums up its reputation. It means 'the swallowing thing who has gone blind' or as we

might say 'the blind glutton'. Wolverine is its other common name. The glutton is for its size (it can reach 75 centimetres in length and about 40 kilograms in weight) one of the most powerful animals known. It was once common over the northern two thirds of North America, Asia and Europe but has been pushed to within the Arctic Circle on all three continents. It is a solitary hunter, dwelling in a burrow during the day and hunting at night. This strong and aggressive animal is known to challenge much larger carnivores, such as bears or wolves, for their kill, and it usually succeeds in taking over the carrion.

Below: *The wolverine is a solitary hunter of the Arctic circle. Fearless when it meets a bear or a wolf with a kill, it will often succeed in taking over the carrion. Eating anything edible, its other name is glutton.*

Honey-guide

Honey-badger

Honey-guides and honey-badgers – hunting co-operation

The honey-badger of tropical Africa and Asia is more commonly called a ratel these days, since all badgers love honey. The African ratel (*Mellivora capensis*) is a thick-set squat animal with small eyes and ears. The back and top of the head is a pale silvery grey, and the lower half of the body is black. It lives in overgrown and forested country, usually alone, but a remarkable association has been formed with the indicator bird or honey-guide (*Indicator indicator*). When a honey-guide en-

counters a ratel it chatters and calls loudly and peristently. Apparently the ratel is attracted towards the source of the noise, making strange hissing and chuckling sounds in reply. The bird keeps moving and calling until it reaches a bee's nest. Here, the ratel takes over, and with its powerful claws rips the nest open and gorges itself on the honey. The skin of this mammal is remarkably tough and forms a useful armour against bee stings (and sometimes snakes' fangs). The bird is now able to obtain a share in the feast; it would have been quite unable to break into the nest itself.

The ratel also eats small mammals, ants, beetles and reptiles, including the deadly mamba, although its preferred diet is honey.

Badgers

There are some 8 species of badger to be found around the world but the best known are the Eurasian badger (*Meles meles*) and the American badger (*Taxidea taxus*). The common Eurasian badger measures up to a metre in length and weighs about 13 kilograms. It is quite a large carnivore and has survived very well in densely populated western Europe. This is mainly because of its secretive, nocturnal habits and because of its diet being mixed rather than consisting entirely of meat.

During the day badgers sleep in their sett, either alone, or more usually within a family group. They are noisy sleepers in their underground home, and snores can be heard through an entrance hole. At night they search for grubs, larvae and worms, as well as frogs, small mammals and fallen fruit and berries. Their sense of smell is very well developed. They do not hibernate in winter as many people believe, although they do spend most of their time dozing, emerging only on milder nights. They do, however, lose considerable weight over the winter months, as food is quite scarce.

After a long gestation period of about 7 months, baby badgers are born in February or the beginning of March. The 2 to babies are blind for the first month and do not emerge from the sett until they are 6 to 8 weeks old. They usually stay with the family throughout the following winter, before setting off to find a home of their own.

The American badger is slightly smaller than the Eurasian badger and tends to live on open sandy plains. The dark brown face has a white stripe down the middle like the Eurasian species but, unlike its relative, this band continues all the way down the back.

Skunks

The black and white skunks are renowned for the ability to squirt, from a distance of several metres, an evil-smelling liquid all over an enemy, such as a puma or a bobcat, and during the ensuing commotion to disappear. Skunks inhabit America, the striped skunk being the most common species in North America, while the hooded skunk replaces it in Central America. Another species, the hog-nosed skunk of Central and South America, has a pink, naked, pig-like snout. Skunks are mainly abroad at night, snuffling around in search of insects and small animals.

Right: *A young badger leaves its sett at dusk to search for food. It will rely mainly on scent and touch as it forages around in the earth for worms, beetles and fruit.*
Below: *The striped skunk* (Mephilis mephitis) *is to be found from the Hudson Bay area to Mexico. It is about the size of a domestic cat.*

An otter
swimming
underwater

Above: *A pair of otters concentrate on something in the water, perhaps a fish or a frog, which they will attempt to catch. River otters are probably the most playful of mustelids, sliding and tumbling around.*

Otters

Of all the mustelids the otters are the best adapted for living in water, and spend much of their time there, the sea-otter being completely aquatic. With the exception of Australia, otters are found all over the world, and all are very much alike. They have lithe, close-furred, streamlined odies with a powerful tail, short limbs and usually broadly webbed hind feet.

The common otter (*Lutra lutra*) of Eurasia is typical of the 20 or more species. It feeds chiefly on fishes but other small aquatic animals may also be eaten. It makes its den or holt in the river bank with the opening often under the roots of trees growing along the bank. This otter, like most others, is mainly nocturnal but the young are sometimes seen playing during the day. They even slide down muddy banks and flop into the water in a similar fashion to humans sliding down water chutes into swimming baths.

The North American otter (*L. canadensis*) is slightly larger than the Eurasian species, which reaches 90 centimetres in length. One third of its length is tail. The largest species is the giant otter or saro of the South American rivers. It can measure over 2 metres, but it is apparently gentle and sometimes young cubs are tamed by tribesmen along the banks of the Xingu river in Brazil. Many of the otters are in danger of extinction because they are hunted for sport and their rich pelage.

The sea otter – a tool-user

The sea-otter (*Enhydra lutris*) is the only otter which lives entirely at sea. In the last few years it has been saved from certain extinction. Now protected by the United States and Russian governments, it can no longer be hunted for its thick, dark brown fur. Today the killer whale is the otter's most dangerous enemy.

Although some sources state that a female gives birth at sea, the more widely held view is that she comes ashore to have her baby, usually during the more clement summer months. The cub (rarely twins) is born in a more advanced state of development than any other mustelid, being well-furred, with eyes open, and with a full set of milk teeth. It is soon in the water being cradled on its mother's tummy as she swims or floats on her back. The mother dives down to the seabed to search for food, such as clams, sea-urchins, other molluscs, crabs, fishes and seaweed. The baby floats at the surface, bobbing helplessly up and down, or being given some support by floating seaweed, until mother

Above: *A sea-otter floats on its back while it smashes a clam on a rock on its belly.*

resurfaces. After 2 to 3 months the cub can swim and dive for itself but it will remain with mother for its first year or two.

The most remarkable feature of the sea-otter is its tool-using ability. While floating on its back, this otter will put a flat rock, that it has selected from the seabed, on its chest. It then pounds certain food objects, such as clams or sea urchins, on the stone until it has cracked them open and can obtain the soft flesh.

Left: *A North American raccoon consumes its fish prey, holding it in its forepaws in a squirrel-like manner. It feeds on a wide range of food from aquatic animals to land snails, birds, their eggs, and fruit.*

America in small bands, numbering between 5 and 12 individuals. They are omnivorous in diet with their snout being very important for finding food. The related cacomistle is more lightly built than a raccoon and is often called the ring-tailed cat, or even the cunning cat-squirrel. It has delicate yellowish grey fur with a beautifully ringed tail. Although agile and active, it is rarely seen, as it is shy and hunts by night.

Of altogether different appearance, mien and habits from the procyonids already mentioned are the kinkajous. The kinkajou

Raccoon family

The raccoons, coatis and kinkajous, are members of the family Procyonidae which is the New World equivalent of the Old World family Viverridae, which contains the mongooses, genets and civets. The only procyonid representatives in the Old World are the red and giant pandas.

The common North American raccoon (*Procyon lotor*) is widely distributed over North America. It is even becoming a suburban garbage-can hunter, and in some places lives a semi-domesticated life, relying on human offerings. Young 'coons' are often tamed but they remain unpredictable in temper and behaviour. The North American raccoon is easily recognized by its bushy black and white ringed tail and its black 'mask'. Its body is covered with longish, coarse black to brown hair. The raccoon usually lives near water and it is here that they search for frogs, water snails, crabs, mussels, and small fishes. On land they eat land snails, earthworms, insects, small rodents, and small

birds and their eggs. Also various plants, including their roots and fruits, are eaten.

Coatimundis are elongated raccoons with a snout which can be upturned to an angle of some forty-five degrees, so that it is almost like a trunk. Coatis roam the forests of Central and South

Above: *The coatimundi, of Central and South America, forages in trees and on the ground for food using the tail as a balancing organ.*

(*Potos flavus*) has a blunt head rounded ears and large eyes These eyes allow it to see wel during the hours of darknes when it is most active. Its legs are short and thick and the slender body has a long curling tail which is prehensile and acts as a fifth limb, when the mammal climbs. The kinkajou and the binturong (a viverrid of south-east Asia) are the only two members of the Carnivora to have prehensile tails. The kinkajou feeds mainly on soft fruit but will eat insects, fungi, nuts and leaves. It relishes honey and is often referred to as a honey bear.

A member of the raccoon family that looks more like a bear is the well-loved giant panda (*Ailuropoda melanoleuca*) of the bamboo forests of western China and Tibet. Although it is very rare, this animal is well known to almost everyone due to the publicity given to the few exhibits in western zoos. Up to 180 centimetres long and weighing 90 to 140 kilograms when adult, this striking black and white carnivore does not eat meat but is specialized for a diet of mainly

bamboo shoots. It sits on its haunches and feeds the shoots into its mouth with its forefeet.

The lesser or red panda (*Ailurus fulgens*) lives high up in the mountains of Nepal, the Himalayas and China. It has very attractive red fur and an attractive cat-like face. This panda is the giant panda's nearest rela-

Above: *The red panda, close relative to the giant panda, inhabits the bamboo forests of the higher altitudes of mountains in the Himalayas, and China. Nocturnal, it sleeps in a tree, curled up like a cat.*

tive, although it is much smaller. They live in pairs with their offspring, and spend much time in the trees.

The giant panda

The giant panda was introduced to the western world in 1869 by the studious French priest Père Armand David, who first saw the striking black and white skin. Père David decided that it was probably related to the bears, but in Paris, Professor Alfonse Milne-Edwards looked closely at the skins and skeletons and said that it must be related to the red panda, discovered earlier in the same century, and therefore be a member of the raccoon family.

Panda country is inhospitable and inaccessible, being the bamboo forests of western Szechwan and eastern Sikang. Here steep-sided mountains are thickly covered with dripping forest. The giant pandas live between 1,500 and 3,000 metres within the limits of the bamboo forests. Most of the giant pandas in captivity are to be found in China and they have been bred there. With better relationships existing between China and western countries there are now several pairs to be found in western countries, as a result of gifts to the heads of the countries concerned.

Chinese zoologists have studied this attractive mammal in the wild and have learned a great deal about its behaviour. It lives alone, a female usually being accompanied by her offspring, however. Although it spends most of its time on the ground it will climb trees with great agility.

It spends about 12 hours a day feeding on the shoots and stems of bamboo. To help hold the stems, the panda has extra pads on its forepaws which are used as 'sixth fingers', opposable to the claws.

Bears

The bears of the family Ursidae are large animals that are instantly recognizable to all. To the zoologist, however, bears are just gigantic, tailless dogs. The bears are all built on the same plan. The hind legs are usually shorter than the forelegs so that the back slopes down from the huge shoulders. All the paws have 5 toes which have large, strong claws. The bears all walk on the soles of their feet (called plantigrade locomotion) which, except in the case of the polar bear are naked. Bears are found in Europe, Asia, and America, but not in Africa or Australia.

The familiar brown bear (*Ursus arctos*) was formerly found over much of western Europe's forested areas, but, due to persecution by man, it is now found only in the Pyrenees and Scandinavia. Brown bears are still reasonably common in eastern Europe, northern and central Asia and North America, however. The common name is somewhat misleading, as the thick coat may vary considerably in colour from pale yellow through reddish brown to black. They also vary in size from the small Syrian brown bear to the giant races from Alaska that are termed grizzly and kodiak bears.

Brown bears are omnivorous, eating all kinds of plants and their fruits and berries, honey from wild bees' nests, fishes and carrion. They may hunt live animals, but as they are quite heavy and have rather an ambling gait they cannot rely on capturing fresh meat. They do not hug their prey to death as old reports state, but rather kill it with powerful blows from the forelimbs and with their teeth. Brown bears can be very dangerous to man and many attacks have been reported. Quite often in American National Parks foolhardy people leave their cars to approach the seemingly tame bear, and deaths have occurred in this way.

The American black bear (*U. americanus*) also eats a wide variety of foods and is still reasonably numerous. However they are sometimes difficult to tell apart, as black bears can be brown in colour. A black bear spends much of the winter sleeping in a warm, dry, sheltered den. The pregnant female gives birth to her cubs,

Below: *A North American brown bear enjoys an annual meal of fresh salmon when these fish return to the headstreams of rivers to spawn. At other times of the year this omnivore will feed on anything from fruit to carrion, also killing prey.*

usually twins, in the den towards the end of winter.

The Asiatic black bear is smaller and is more often known as the moon bear, since it has a broad crescent of white hair on its chest. The smallest, and only tropical bear, is the sun bear of the tropical forests of south-east Asia, Sumatra and Borneo. The yellow crescent on its chest is

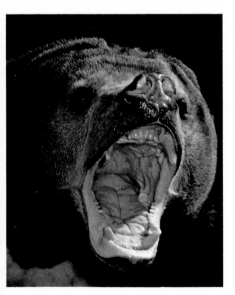

Above: *The sun bear* (Helarctos malayanus). *This is the smallest and the only tropical bear of the family, inhabiting the jungles of south-east Asia. It feeds on wild bees and termites.*

said, in folklore, to represent the rising sun and this is how it got its name. This bandy-legged, small, light bear is an expert climber, searching the trees for lizards, birds, fruit and bees' nests. The sloth bear of the forests of southern India and Ceylon also has a crescent on its chest. It looks rather scruffy because of its long and shaggy fur and it got its name from its slow, shuffling walk.

The only true carnivore among the bears is the polar bear (*Thalarctos maritimus*), which lives on the inhospitable shores of the Arctic Ocean. It is not surprising that it eats mainly meat as there is little else available. In late summer they can eat ripe berries from the tundra's plants. However, seals, especially the ringed seal, make up the bulk of the polar bear's food. Fishes, and seabirds and their eggs are also taken when available. When food is scarce, this bear will roam great distances searching for food, and will scavenge on remains of walrus, left by hunters, or eat stranded whale carcasses.

The polar bear's white coat helps to camouflage the animal when it is stalking seals across

Above: *Polar bears are the most carnivorous of all the bears, since within the Arctic circle there is little else but fresh meat available. They will hunt seal, fishes, sea birds and their eggs.*

the ice or when it is lying in wait at a seal's breathing-hole in the ice. A seal, on poking its head out of the water, is dragged on to the ice with the bear's long claws and then is delivered a stunning blow.

Polar bears lead solitary lives, pairing briefly for mating and then parting to lead their own lives once more. During the harsh Arctic winter, males tend to move southwards to less hostile regions, while each pregnant female constructs a den beneath the snow. Here in late November or December, she gives birth, usually to twins. Like all bears at birth, they are tiny, blind, and naked. The mother suckles them and cares for them, showing much devotion. The cubs do not venture from the den until the spring. They stay with their mother for at least a year, learning hunting techniques from her. They are very playful during their first summer, often being joined in their games by their attentive mother.

Dog family

The family Canidae includes wild dogs and domestic dogs, wolves, foxes, jackals and coyotes. They all have long legs and walk on the toes of their rather small feet (digitigrade locomotion). The hind foot of all dogs has only 4 toes. A canid's head has a long jaw and this gives the dogs and their relatives the typical long-muzzled appearance. A dog's tail is not as long as that of many cats and is used for signalling the dog's emotions. Dogs are unable to climb trees but they are excellent runners. Most dogs are flesh-eaters and all possess long canine teeth or fangs which are used for piercing and tearing their prey. Large shearing cheek teeth are used to slice the meat into swallowable chunks. When hunting, dogs rely mainly on their excellent sense of smell, helped by their good hearing: sight is less important until the canid comes close to its quarry.

Wolves

The wolf (*Canis lupus*), a close relative of the domestic dog, once ranged across Europe, Asia and North America, in the forested areas. It has largely been exterminated in the temperate areas and survives only in the colder northern parts of the northern hemisphere. The timber wolf, as it is often called, looks rather like the domestic alsatian in size and shape. To the south, in North America, it is replaced by the coyote or prairie wolf (*C. latrans*), which is smaller and hunts small mammals and birds as well as scavenging when it can. It inhabits the open prairies, although recently agriculture has driven it into more mountainous areas. In Asia, the wild dog (*Cuon alpinus*), known as the dhole in India, replaces the timber wolf. This reddish-brown animal hunts in packs and will attack large mammals, such as wild boar and spotted deer, and it has been known to kill and eat sloth bears and black

bears. One report states that a tiger was pulled down, when fleeing from a tree refuge, and killed by a vicious pack.

Jackals

Jackals are found in Africa and Asia, and are closely related to the wolf and the domestic dog. They can, in fact, interbreed with them. They look like small wolves and hunt mainly by night, always alert for their main enemy, the leopard. Mostly they hunt alone, killing rats, mice, birds, reptiles, and even small antelopes. Packs are formed when there is a chance of sharing the remains of a lion or leopard kill. The four species of jackal are the black-backed, side-striped, simenian, and golden.

Above: *Timber wolves feast on an American elk. Once distributed across North America, Europe and Asia wolves were wiped out from all settled areas, surviving in wilder uninhabited areas. Within the Arctic circle most of the wolves are white all the year round. They are social animals, hunting together in small family packs.*

Foxes

Foxes are night-time hunters, resting or sleeping by day in their burrows or 'earths'. The red fox (*Vulpes vulpes*) of North America, Eurasia and North Africa is a sly and cunning hunter, never collecting in packs, although a vixen and a dog fox sometimes hunt together. The red fox has survived better than its relatives the changes man has made to the Eurasian and North American landscape. In many parts of its range, this fox is found within the boundaries of towns and cities, surviving by scavenging from dust-bins and refuse tips. The usual diet is any prey from insects to mice and rabbits. It is a fact that some develop a taste for domestic poultry, although the numbers are not high.

The pale, cream-coloured fennec fox (*Fennecus zerda*) of the deserts of North Africa and Arabia is most attractive, with its fluffy coat and huge ears. These ears can pick up the minutest

Above: *A red fox of Europe and Asia is often called cunning, due to its intelligence and keen senses of smell, sight, and hearing, outwitting man who has relentlessly hunted it through the ages.*

sound, which helps the fox to locate its prey at night. Another attractive fox is the arctic fox (*Alopex lagopus*) of the tundra lands within the Arctic Circle. In summer, this fox has a short-furred, grey-brown coat which is replaced by a long, thick, white coat for the snowy, harsh, winter months. Some races have blue or silver coats in winter. The diet is mainly carnivorous, comprising seabirds, ptarmigans, fishes, arctic hares, and seal pups. Lemmings make up a high proportion of this fox's food in summer, and when lemmings have a population explosion, every 3 to 5 years, the number of arctic foxes increases also. When food is very scarce, during winter, the arctic fox often follows hunting polar bears in the hope of getting the remains of a kill.

Above: *The attractive fennec fox lives in arid and desert regions of northern Africa and Arabia, resting in a burrow by day and being active at night. The huge ears help to cool the animal by radiating heat as well as picking up the sounds of prey.*

Seals and their relatives

(ORDER PINNIPEDIA)

The seals are virtually world-wide in distribution, 1 or more of the 32 different species being found along almost all coastal regions, including those of both the Arctic and Antarctic. All the species are carnivorous and in the past many have classed the seals and their relatives as a suborder of the order Carnivora. Today they are given their own order, the Pinnipedia. This name means literally 'fin-footed' or 'feather-footed', and it refers to the fact that the limbs have evol-

Below: *Grey seal mothers with well-developed pups. The young have already lost their white, woolly coats and soon they will depart from the beach with their mothers, returning to shore the following year for the next breeding season.*

ved into swimming flippers as an adaptation to a marine life. All members of the family are streamlined, this torpedo-shape or fish-shape enabling them to move through the water very swiftly. Unlike the whales and the sea-cows, seals have retained their hind limbs which, although flipper-shaped for swimming, are some help with movement when the animal comes on to dry land for breeding each year. In the fur seals and sea-lions, the hind limbs can actually be rotated forward to help with movement on land. Most seals eat various kinds of fishes, but other sea animals, such as molluscs, crustaceans, and sometimes sea birds, such as penguins, may be taken by some species.

The nostrils and ear-holes are closed when the seal is diving and, like all aquatic mammals, it must surface every few minutes to breathe. The deepest diving seal is the Weddell which has been recorded, with a depth gauge, diving to 600 metres. When asleep in the water, a seal slowly rises to the surface still fast asleep, breathes, and slowly sinks again. This can be seen in captive specimens in zoos. The order is divided into 3 families; the fur seals and sea-lions, the walrus, and the true seals.

True seals

The true seals include the common, ringed, harp, grey and monk seals as well as the feared leopard seal and the gigantic elephant seal. They all have short necks and no external ears, the openings being found on the sides of the head. These ear-holes can be closed as the mammal submerges. When swimming, seals hold the webbed hind feet close together and move them from side to side so that they somewhat resemble the tail fin of a fish. The front flippers are used to steer the

Californian sea-lion (*Zalophus californianus*) – an eared seal

Common or harbour seal (*Phoca vitulina*)

animal, although occasionally they help in movement.

On land, true seals are extremely clumsy. Their hind limbs are unable to be moved under the body to help support them and the front flippers are quite short. So the seals have to drag themselves up the breeding beaches by hunching up the back, and throwing the head and front of the body forwards, and then dragging the hind part along.

One of the best known of the 18 species of true seal is the grey seal (*Halichoerus grypus*). It is found on the coasts of northern Europe, Iceland and Greenland and the majority (over half the world population) gather in 'rookeries' for breeding on the rocky coasts

of the British Isles. The young pups are born and then the females are mated by the bull before they return to the sea. The common seal (*Phoca vitulina*) is found on both sides of the Atlantic and its pups are born usually towards the end of June. The ringed seal (*Pusa hispida*) is the smallest of all the seals and the adults have a pale ring around the dark spots. Although found in North Polar seas, they are most common off Greenland. It is thought that the Baikal seal (*P. sibirica*), the fresh-water seal from Lake Baikal, originated from the ringed seal. It evolved separately when Lake Baikal was cut off from the Arctic Ocean many thousands of years ago.

The most ferocious seal is the leopard seal (*Hydrurga leptonyx*) of the antarctic seas. It attains a length of up to 4 metres and has exceptionally long, sharp, cusped teeth, well designed to pierce and tear the flesh of prey. It is the chief predator of the antarctic penguins and often lurks in the water around ice floes or coasts where colonies are present. They also feed on cuttlefish, fishes, seabirds and carrion, often that thrown overboard by whaling ships.

The elephant seal (*Mirounga angustirostris*) is the giant among the seals, a fully grown bull measuring over 6 metres and weighing nearly 4 tonnes. This is just a little less than the weight of an ordinary elephant, and so the seal is well named. The trunk-like snout of the adult bull also names the species. In the breeding season, when the bull attracts females and sees off bulls intruding

Right: *An elephant seal roaring, his bulbous snout inflated to warn off an intruder. Fully grown he can measure over 6 metres.*

into his territory, the nose is inflated to over 60 centimetres. There are two distinct forms, the northern and the southern elephant seals, the southern ones being much more common.

Walrus

The walrus (*Odobenus rosmarus*) is the only member of its family and it lives in the shallow waters at the edge of the polar ice in the Atlantic and Pacific Oceans. Like the eared seals the walrus can rotate its hind limbs but like the true seals it has no external ears. These animals can reach over 3 metres in length and both sexes have long tusks or canines growing down from the upper jaw. They are used for defence and for digging shellfish, starfish and

crustaceans up from the sea bottom. The coarse white whiskers act as organs of touch.

Below: *Walrus colony. The long tusks are used to gain a hold when the animals haul themselves out of the water.*

Ruthless slaughter of seal pups

The slaughter of young white-coated pups has aroused great public opposition: perhaps only whaling commands more publicity and sympathy. The whitecoats, as the babies are known, are battered to death with baseball bats and then skinned. Many reports state that the defenceless pup may still be alive while being skinned, however. Recent findings show that the harp seal of the north-west Atlantic has been greatly over-exploited and the mass slaughtering is still taking place. There are fixed quotas now set, but these often account for the whole year's young seal population. Although governments and international bodies are taking action, commercial sealing is still a threat to most seal species.

Eared seals

The eared seals as their name suggests have small external ears which identify them. They have a longer neck than true seals and the webbed hind feet can be rotated under the body to help with movement on the land. The Californian sea-lion is the best known, since it is the one usually seen in zoos or circuses. In the wild it is extremely agile and feeds mainly on squid. The other eared seals are the fur seals. The old bulls have a thick mane round their neck and roar like a lion during the breeding season.

Below: *Californian sea-lions performing underwater acrobatics.*

Elephants

(ORDER PROBOSCIDEA)

The elephant is the largest living land animal. There are 2 species, the African (*Loxodonta africana*) and the Indian or Asian (*Elephas maximus*). There are two types of African elephant, a smaller rain forest form, confined to the dense jungles of West Africa and a large bush form which is widely distributed over tropical Africa. The Indian elephant has recognized races in Bengal, Ceylon, Sumatra and Malaya. In all these areas the Indian elephant has been domesticated for centuries, and it helps man with the timber work in the forests.

From fossil evidence it is known that many species used to live on the Earth, ranging all over it with the exception of Australia and Antarctica. Some of these elephants – the mammoths – survived long enough to co-exist with the early races of man, and complete mammoth specimens have been found, frozen for thousands of years, in northern ice fields.

The African and Indian elephants are similar in physical appearance and structure. They have a massive body, large head, short neck and stout pillar-like legs. The feet are short and broad with the digits of the fingers and toes ending in hoof-like nails. These surround the cushioned pad of the foot's sole.

The outstanding feature of an elephant is its trunk, which is formed from the nose and upper lip, with the nostrils placed at the tip. This very flexible snout is used for breathing and smelling, for carrying food up to the mouth,

Left: *Drinking time for a herd of African elephants. Water is drawn up into the long mobile trunk and then squirted into the mouth or over the body for a cooling shower.*

and for sucking up water to eject into the mouth or to spray over the body when bathing. It is also used for spraying dust and fine soil over the animal when dust-bathing, as well as for grasping, pulling and lifting various objects. At the tip of the trunk are finger-like projections which are used to pick up small objects, the African species having 2 'fingers', and the Indian having only 1 'finger'.

Above: *An Asiatic elephant with its master, the mahout. This species has been domesticated for centuries, being an intelligent, docile worker when well-treated.*

The single incisor teeth, one on either side of the upper jaw, grow into elongated curved tusks which can be extremely long in old African males. The huge molar teeth appear in pairs in the jaws, one tooth on each side of the mouth. As one is worn down, by

Indian elephant
(*Elephas maximus*)

Lives in shady forests

Smaller ears

African elephant
(*Loxodonta africana*)

Lives in hot, dry plains

Large ears which are fanned to keep it cool

Tip of trunk has one finger

Tip of trunk has two fingers

Hyraxes – small, distant relatives of the elephant

Hyraxes (*Hyrax* species) are referred to in the Bible as conies and they are also known as dassies. They are placed in their own order, the Hyracoidea, which contains 6 species, all of which are found living in Africa. The tree hyraxes are found in forested areas, while the ground hyraxes live mainly in rocky areas. They look rather like rabbits or compact rodents, at a distance. There are definite features linking the small hyrax, at 3 kilograms, with the massive elephant, at 6 tonnes, but zoologists are cautious when it comes to tracing a direct link. Embryology and the reproductive organs show certain likenesses. Extinct hyraxes were probably larger than the modern elephants!

Hyraxes are very agile when scrambling over rocks or when climbing trees. The secret of their climbing skills lies in the pads on the soles of their feet. They are comparable to the soles of a climbing boot, both giving a firm grip on smooth, slippery surfaces, even when steeply inclined. After a morning toilet, the rock hyraxes nibble grasses or available vegetation, or sunbathe on the rocks. They always keep an alert watch for the snakes, leopards, servals and jackals that are their main natural enemies.

Right: *Rock hyrax enjoying a sunbathe. The soles of the feet have non-slip pads on them which help them grip the rocks.*

the continual grinding of the plant food, it is replaced with a new one growing below it. The old molar eventually drops out.

The main differences between the species are the much larger ears and longer tusks of the African elephant. The ears not only give excellent hearing but are waved back and forth to help cool the animal when it gets too hot. The African elephant also has a sloping forehead and hollow back, while the Indian species has 2 domes on its forehead and a convex curve to the back. The African elephant is the much larger animal, an adult bull growing up to a height of 3.5 metres and weighing up to 6.5 tonnes.

The elephant is entirely herbivorous, and as plant material is low in food value and the animal is so large, it has to spend about 18 hours each day feeding to produce enough energy to survive. It rests only in the middle of the day and for part of the night. The trunk gathers leaves, shoots, bamboo, reeds, grasses and fruits, as well as cultivated crops such as maize and bananas. The elephant eats about 5 per cent of its own weight a day and drinks about 180 litres of water, sucking up 9 litres at a time with its trunk.

Family life

Elephants usually live in well-organized herds, ranging in size from a small family group of 4 to 6 individuals to a large herd of between 20 and 30 animals. Sometimes the herd is led by a mature bull, but often an elderly, experienced cow will lead it. In a large herd, besides the leader there will be a few young bulls, a number of cows and their offspring. Old bulls often lead a solitary life.

An African cow elephant gives birth to a single baby after about 21 months gestation. The newborn calf immediately struggles to its feet and reaches for its mother's nipples, situated between her forelegs. It suckles with its mouth while the trunk lolls away to one side. As the calf grows it learns to use its trunk for the different functions. The mother shows great attention to her offspring; she constantly caresses the little calf with her trunk, cleaning all its folds of skin and its openings. If she suspects any danger she pushes the calf under her belly and if they are on the move the calf runs along under here. As it gets older and too big to do this it will run at her side or hold on to her tail. The calf is only fully weaned at about 5 years old and does not mature until it is about 15. The elephant's lifespan is very similar to man's, the average being 60 or 70 years.

The African elephant has few enemies apart from man and leads a peaceful life, unless upset by an occasional aggressive lion, or by an intruder such as an unwary observing naturalist, or perhaps by a low-flying light aircraft. The leader of the herd will raise its trunk and test the air to see if it can smell out the danger. An urgent blast is given through its 'trumpet', telling the herd to keep together and gather their offspring to them. If the herd leader detects the source of danger visually, it will roll down its trunk, throw back its ears, raise its head and probably charge the intruder. If the rest of the herd is to follow, it is told to advance by screams or growls from the leader.

Below: *An African elephant mother shelters her calf under her belly. The calves are protected in mixed herds which are led by a senior female. In the rainy season she will lead the herd over long distances to new feeding grounds.*

Left: *Underwater shot of a manatee. This sluggish, generally inoffensive creature can stay submerged for about 16 minutes, although it generally will surface for 2 or 3 breaths at 5 to 10 minute intervals. In shallow water it will 'walk' on the inturned tips of the hind flippers.*

Sea-cows

(ORDER SIRENIA)

The sea-cows or sirenians form a group of aquatic mammals that includes the modern dugongs and manatees and the extinct Steller's sea-cow. Dugongs are found in the coastal waters of the tropical areas of the Old World, and occasionally they enter estuaries and travel up rivers. Manatees live along the coasts and in coastal rivers of the south-eastern United States, the West Indies, northern South America, and western Africa. Steller's sea-cow became extinct in the eighteenth century, but formerly lived in the cold waters of the Bering Sea. Man's greed and natural causes brought about its swift extermination.

Sirenians are massive, spindle-shaped animals with paddle-like forelimbs, no hind limbs and a whale-like tail. The adults range from 2.5 to 4 metres in length and weigh up to about 360 kilograms. The rounded head has a flattened muzzle, and the small mouth is surrounded by thick, stiff vibrissae (tactile hairs). The nostrils can be closed so that the breathing system can be cut off when the sirenian submerges.

Dugongs and manatees differ in several points. In dugongs the tail fin is deeply notched, but in manatees, it is more or less rounded. The male dugongs grow tusk-like incisors, but these are not present in manatees. Manatees have up to 10 cheek teeth in each half of the jaw, many more than dugongs have. These are also replaced consecutively from the rear, an action similar to the replacement of the huge cheek teeth of elephants.

Dugongs are seen either alone, in pairs, or in small groups. They appear to be quite affectionate towards each other. They feed underwater on various seaweeds and marine grasses. The mammal

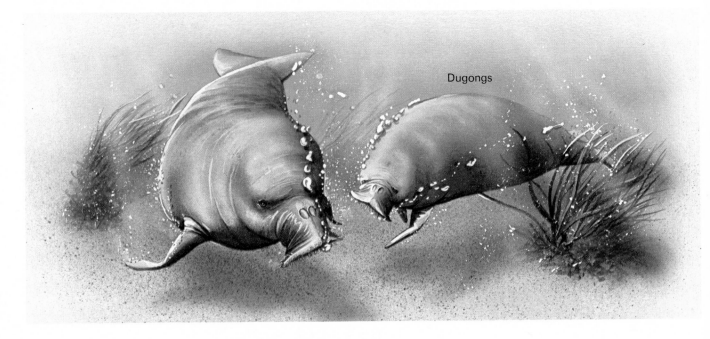

Dugongs

Dugongs – the sirens of mythology

Dugongs are probably the sirens of mythology, for some of the stories of mermaids are undoubtedly based on these aquatic mammals. They were first mentioned in western literature by the Greek philosopher Pliny. They may also have been the basis of Homer's Greek myth of Odysseus and the sirens. The early European seafarers brought back tales of seeing mermaids nursing their babies at the breast like a human mother breast-feeding her baby. They no doubt did see the peaceful browsing dugongs at sea: females do nurse their babies at twin nipples situated in the chest region. However, nursing usually takes place underwater as it does with the manatees.

Dugong nursing its young

A female gives birth to a single baby after about 11 months of development and it appears that breeding can take place all the year round. The mother is most caring with her baby and guards it well. Another aspect of behaviour that might have given rise to mermaid myths is that she will sometimes carry her baby on her back. Seamen without the help of spectacles, but with their daily tot of rum and long absence from home, can certainly be forgiven for seeing fair maidens in the sea.

Above: *An illustration from 'The Little Mermaid' by Hans Christian Anderson. The mermaids were the sirens of mythology and this name was given to the order of dugongs, manatees and seacows, the Sirenia, because of the mermaid-like manner of nursing in these mammals.*

rips out the entire plant, swishes it backwards and forwards in the water until most of the sand is removed and then swallows it, with little chewing. They also pile sea grass in stacks along the shore and eat it at leisure, later.

Manatees gather in much larger groups of 15 to 20 individuals when the young are half-grown, but they are usually found alone or in small family groups. The large groups of half-grown individuals form because they are too old to remain in the family, but are still too young to find mates. In Florida they are known to migrate to warmer spots in cold weather and there congregate in even larger numbers. They often greet one another with 'muzzle-to-muzzle kisses'. Like the dugongs they are strictly herbivorous but will take marine and freshwater plants.

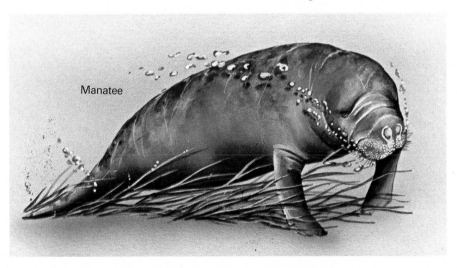

Manatee

Hares and rabbits

(ORDER LAGOMORPHA)

The hares, rabbits and pikas of the order Lagomorpha are found almost all over the world. They are only absent in the Antarctic. Although they do not naturally occur in the Australian–New Zealand regions, man has all too successfully introduced them there.

The pikas or conies (*Ochotona* species) are basically rock-dwellers, 12 species living in Asia and 2 in North America. Most of them occur in the colder, more northerly and more mountainous regions. Pikas have short, rounded ears and rounded bodies and they live in burrows under rocky outcrops. They have many amusing local names such as rock

Left: *Mad March Hares, the term used when these animals rise up on their hind legs, spar with partners and chase over the fields.*

Left: *Eurasian rabbits feeding in the late afternoon sun, still quite close to the safe retreat holes of their warren.*

rabbit, little chief hare, mouse hare, haymaker, squeak rabbit and whistling hare. To survive the winter they dry grasses in summer and store them under overhanging rocks as 'haystacks'. These act as winter restaurants for the pikas when food is otherwise unobtainable.

The 52 species of rabbits and hares are found from the freezing Arctic Circle to the shimmering heat of the deserts. Their ears and limbs are longer than those of pikas and they are extremely fast moving. All have short tails. Some are good at burrowing, whereas others keep to the surface, never digging. They all have large, chisel-like, ever-growing front teeth, the incisors, and a gap, or diastema, between the front teeth and the cheek teeth. In

this respect they are similar to rodents and, at one time, until it was realized that the superficial resemblances were due to a similar way of life, they were included in the rodent order.

The European brown hare (*Lepus europaeus*) is larger and lankier than the European rabbit (*Oryctolagus cuniculus*) with longer, black-tipped ears and longer hind legs. When running, the stilt-like action of the hare distinguishes it from the bobbing gait of the rabbit. It is mainly solitary, although in early spring bands may be seen chasing and 'boxing' one another, in open fields, as a preparation to mating. It is not certain whether it is just the bucks, or both sexes that box. The European hare does not burrow but rests in a form, a hollow in coarse grass. Litters of 2 to 4 young are born in an advanced state of development, with their eyes open, fully furred and able to

use their limbs immediately.

The European rabbit lives in extensive burrows or warrens, although in some parts of the range, where myxomatosis wiped out a large percentage of the population, the surviving rabbits and later generations have taken to living above ground. Between 3 and 12 young are born to the doe on a bed of grass, dried in an underground nursery chamber. The doe does not allow a buck near and tends the blind, naked rabbits by herself. They can look after themselves within a month and by then the doe may be pregnant again, giving birth to another litter some 2 weeks later.

Other species include the arctic hare (*L. arcticus*) and the blue or varying hare (*L. timidus*) both of which turn white in winter in their arctic or mountainous homes. There are various species of jack rabbits and cotton-tails in North and Central America.

Angora

Dutch

Chequered Giant

Domestic rabbits

Domestic rabbits have been bred from the wild rabbit and there are over 50 different breeds recognized today. Among the largest varieties are the Flemish Giants which may weigh more than 6 kilograms. In the Lop breed the ears are very long and droop down to touch the ground. The medium-sized Angora rabbit is usually white and most attractive with its long, thick fur making it look very fluffy. The black and white Dutch rabbit is often kept as a pet by children, as it is hardy and friendly. The Ermine Rex is a small rabbit, usually an albino with pink-red eyes.

Rodents – the gnawers

(ORDER RODENTIA)

The gnawing rodents are found everywhere. There are over 1,700 species and the numbers of some are enormous. It is said by some people that there are more rodent species than all the other mammal species put together. This is rather an exaggeration but the order Rodentia is certainly the largest of the mammal orders.

All rodents have a single pair of chisel-like incisors, in both upper and lower jaws. They are permanently growing and are kept in check by the rodent continually gnawing hard foodstuffs or other objects. The name rodent comes from the latin verb *rodere* meaning to gnaw. As in the Lagomorpha, the rodents have a gap or diastema between the front teeth and the cheek teeth. The shape of the jaws, structure of the skull and the muscles working the jaws all give strength to the cutting movements. With their great adaptability and gnawing teeth, rodents have filled many niches in the environment that other species have been unable to fill. They burrow, run, climb, hop, glide, leap and swim in environments ranging from desert and tropical jungle to polar waste and mountain top. One thing that has probably helped the rodents to become so successful is that the majority are small-bodied creatures, only a few centimetres in length. Only a handful reach the size of a small pig. There are 3 main groups of rodents, the squirrel-like, the rat-like and the porcupine-like.

Above, right: *The American grey squirrel has successfully invaded Britain and Europe.*
Right: *Flying squirrels really glide through the air on 'parachute' skin membranes that run from front to hind legs.*

Prairie dog towns

The prairie dog or marmot (*Cynomys ludovicianus*) lives in large colonies on the extensive grasslands in central North America. A prairie dog town is a complex of underground burrows, that sometimes extend for several kilometres, with hundreds of entrances. Some of these rodents act as sentries, sitting bolt upright on an entrance mound. As soon as any danger is seen they utter a piercing call and all the prairie dogs above ground dash for the nearest burrow. Their burrows are often used as hibernating quarters by rattlesnakes, and when this happens a prairie dog family will move on and dig another burrow network. Sometimes burrowing owls make use of a burrow, also.

Prairie dogs have been persecuted during this century by farmers as the American prairies were taken over for agriculture. Numbers are now much reduced and massive townships are seldom found.

Left: *Mother marmot and family watch alertly from the rim of an entrance mound to their underworld burrow system.*

Squirrel-like rodents

This group includes not only the squirrels, but the marmots, beavers, gophers and chipmunks. They are distributed throughout the world with the exception of Madagascar, Australia and the polar regions. The grey squirrel (*Sciurus carolinensis*) of eastern North America and Europe (where this has been introduced) and the Eurasian red squirrel (*S. vulgaris*) are typical examples of the successful tree squirrels. They are acrobatic climbers, nesting in tree hollows or in specially built nests called dreys, constructed from twigs and leaves in the fork of a tree. They eat nuts, buds, fruit, berries, insects and even young birds that they have killed.

Flying squirrels are found mainly in southern Asia, but there are a few species in Europe, and Central and North America. The giant flying squirrels of the genus *Petaurista* are almost 1 metre in length and are known as tree dogs. They can glide for over 450 metres as they move from a tree top to the lower branches of another tree. All flying squirrels glide on an extended membrane of skin that runs between the legs and acts as a parachute.

The ground squirrels include the sousliks, gophers, chipmunks, rock squirrels, woodchucks and prairie dogs. Most of them have cheek pouches in which to carry seeds and other food and they live in extensive burrows underground. The prairie dogs are renowned for their townships.

Beavers
The largest of all European rodents is the European beaver (*Castor fiber*). It is very similar in

Below: *A Canadian beaver* (Castor canadensis). *This rodent is a master builder, constructing a dam and house across streams or rivers.*

appearance to the Canadian beaver, and is, in fact, thought today to be the same species. From nose to tail it measures about 1 metre and it is different from all other rodents in having a broad, flat, scaly tail, which is almost hairless. The tail acts as a rudder when the beaver is swimming, as a warning device when it is slapped on the water's surface as a beaver dives on becoming aware of danger, and as a prop when the rodent is gnawing through a tree. The beaver lives in family units of up to 12 or so animals. It is a master builder, constructing dams and lodges in streams or rivers. The chosen spot is usually near to aspen and willow trees which provide its favourite food and building materials.

Rat-like rodents

The large rat family contains the familiar rats, mice, gerbils, hamsters, voles, and lemmings. Rats and mice evolved some 40 million years ago in the northern hemisphere and then spread to the other parts of the world. We can only look at a few individual species from this enormous number of small, rather nondescript rodents. Of the New World mice, the deer-mouse or white-footed mouse of North American woodlands does quite a lot of good by eating millions of the insect population. Both America and Eurasia have their own species of harvest mouse. They construct beautifully woven nests of grass slung between grasses.

The Old World mice include the infamous rats (*Rattus*), mice (*Mus*) and numerous others, such as the dormice, field mice, spiny mice, and giant rats. The brown rat (*Rattus norvegicus*) is probably the most adaptable animal ever evolved. Although originally from the tree-less steppes of central Asia, it started its mass emigrations probably in Roman times or earlier and, hitching rides on ships, it has reached all the corners of the world. The common house mouse is also world wide in distribution due to introduction by man and it has the ability to survive and even breed in the most hostile places. Some have bred in the huge refrigerator plants of food factories, for example.

Dormice

The dormice, whose group includes the common dormouse (*Muscardinus avellanarius*) and the fat or edible dormouse (*Glis glis*), have long tails, sometimes very bushy as in the edible dormouse, and they are all excellent climbers. In the autumn they all hibernate, the heart-beat and respiration rate slowing down and the body temperature falling to just above that of the surroundings. They sleep curled up in a nest of dry grasses and leaves.

Below: *The infamous black rat* (Rattus rattus) *eats anything and destroys far more than it eats. Native to Asia Minor and the Orient this species came to Europe with the Crusades, and to North America on board the ships of early explorers. It harbours and carries diseases such as black plague, typhus and rabies.*

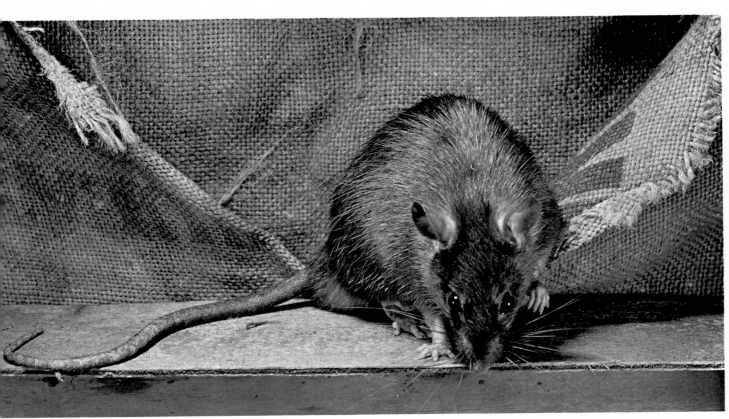

Jerboas

There are some 25 species of jerboa, most of them being confined to Asia with 3 occurring in North Africa. They have enormously elongated hind legs on which they hop rather than walk or run. Although only a few centimetres long they can leap up to 2 metres in a single bound. They are usually sandy coloured so that they blend with their sandy desert surroundings. Jerboas are quite attractive with large eyes and ears, long whiskers, and short front legs which are used for holding food.

Right: *Dormouse in dormancy. In this state of sleep, the heart-beat and respiration rate slows down and the rodent's temperature drops to just above that of its surroundings.*

Lemming migrations

There are many species of lemming and all of them, like most rodents, breed rapidly, so that every so often there is a population 'explosion'. The numbers increase so rapidly that vast hordes of them migrate. In the case of the Norway lemming (*Lemmus lemmus*) 'lemming years' occur every 3 to 5 years. Reports on television and in the press describe huge numbers stopping trains and road traffic as they journey onwards. Some reach the coast where they swim out to sea, continuing until they die from exhaustion. This has led to the popular idea that lemmings commit suicide. It seems that these migrations are triggered off by over-population pressures. They ensure the survival of the remaining individuals, which are then able to produce healthy offspring in the intervening years, before the next population explosion.

Porcupine-like rodents

The two groups of porcupines are the New World porcupines and the Old World porcupines. Those living in America are tree-dwelling animals with sharp claws and rather short quills. The tree porcupines of Central and South America have prehensile tails, while the North American porcupines have rather short, flat tails. The Old World porcupines have very large and characteristic quills on the back which can be erected by the powerful skin muscles. They usually live alone,

Right: *A prehensile-tailed porcupine is lazy and slow in movement but is a sure-footed climber of Central and South American forests. It eats leaves, tender stems and fruits such as bananas.*

Left: *A tree porcupine has highly specialized hands and feet for climbing. The 4 digits of each limb have long, curved claws for excellent gripping of branches.*

sleeping by day in their underground run, and moving slowly around at night searching for food, such as juicy plants, roots, fruit and bark. The porcupine rustles its quills as a first warning to an enemy, such as a dog or wild cat, and if this does not drive the intruder away, then the porcupine turns its back and runs rapidly backwards to the advancing enemy, impaling it on its now stiffly erected spines.

Capybara

The guinea pig, the well-loved family pet, is related to the largest rodent, the capybara or water cavy, of South America. The capybara looks rather guinea-pig-like with a touch of hippopotamus. It weighs over 50 kilo-grams and can be over 1 metre in length. They live in large groups along river banks where they feed on grasses and water plants. At the least sight of danger all the capybaras plunge into the water to escape their enemies, such as jaguars and pumas.

South America is the home of many varied rodents, including the mara or Patagonian cavy, the most hare-like of all the rodents; the spotted pacas; the long-legged agoutis; and the fine furred but now rare chinchillas. The coypu, a large aquatic rodent of South America, is now well established in Europe due to escapes from farms which rear this large rodent for its fur, known as nutria.

Below: *South American coypu* (Myocastor coypus) *resting above its watery habitat. The nipples of the female are placed high on the sides of her belly so the young can nurse while they ride on her back as she swims through the water.*

Ungulates – the hoofed mammals

Most of the species in the two orders of ungulates are familiar, large, fast-moving herbivores. The orders include deer, horses, antelopes, cattle and camels. Their bodies are well adapted for running fast over wide grassy plains or savannas, in order to escape from predators. To do this their limbs are very long and they have raised themselves up so that they move on the very tips of the feet. The toes are sheathed with solid hooves which have evolved from the claws or nails. It is these on which the animal moves, the rest of the foot not touching the ground. This method of locomotion is called *unguligrade*.

The cheek teeth of the hoofed mammals usually have broad crowns, an adaptation for grind-

ing and crushing tough plant food into small particles, so that when it is swallowed the digestive juices can act upon as large a surface as possible.

Ungulates are found wild in all parts of the world, except in the Australasian region. Even here though, camels and donkeys are found living in the hot, dry interior of Australia as a result of escapes from exploration parties.

The odd-toed ungulates or perissodactyls

(ORDER PERISSODACTYLA)

There are only some 16 species of odd-toed ungulates surviving today and they can be divided into 3 groups – the horses which are fast-moving; the tapirs which are shy and retiring; and the rhinoceroses which are large and armoured. The feature that groups these animals together is the fact that the central bone or axis of the foot passes through the middle digit of the toes. This means that these ungulates usually have either 1 or 3 hooves on each foot, and are thus called odd-toed ungulates.

Rhinoceroses

The large, stout body with thick powerful legs, the tough, almost naked skin and the horns on the snout make it easy to identify a rhinoceros. There are 5 species alive today, 3 in Asia and 2 in Africa. Unfortunately, most of them are threatened with extinction.

The largest of the Asian species is the impressive great Indian rhinoceros (*Rhinoceros unicornis*) which lives in the jungles and tall grasslands of Nepal, Bengal and Assam. This huge

Left: *An African black rhinoceros mother stands quietly but on the alert as her calf feeds on her milk.*

Left: *A great Indian rhino is accompanied by a cattle egret that will snap up any insect disturbed by the heavy moving ungulate.*

rhino weighs up to 2 tonnes and stands as much as 1.8 metres high at the shoulder. This is the rhino whose body is covered with heavy 'studded' armour that is folded at the neck, shoulders and leg tops to allow movement. The only hair visible is the tuft on the ears and the tip of the tail. The beast has a single horn on its snout which gives it its specific name *un-icornis*. It has a moveable prehensile upper lip which can grasp tall reeds and grasses on which it feeds. The other species from Asia are the Javan and Sumatran rhinoceroses both of which are few in number.

The largest of all rhinos is the African white rhino (*Diceros simus*). A large male may reach 2 metres at the shoulder and weigh as much as 4 tonnes. Among land mammals only the elephants are heavier. This rhino has a square lip which is an adaptation for its way of feeding: it grazes, cropping the grass with slow deliberate movements. In the African black rhino (*D. bicornis*) the upper lip is long and pointed and forms a prehensile organ which is used to browse on leaves and twigs of trees and bushes. It is the most numerous of all the rhinos. The common names of these African rhinos are rather misleading since they are both dark grey in colour. The name 'white' probably comes from the Dutch word *wijd* which means wide and refers to the shape of the lips. Both the black and white rhinos have two horns which sometimes grow extremely long. Rhino horn is composed of very closely matted hair.

All the rhinos are renowned for their aggressive charges at cars or intruders. Normally they are placid, gentle beasts, and it is probably their weak eyesight that makes them nervous and aggressive towards any sudden nearby movement. Their senses of hearing and smell are very acute and are relied on more than their sight.

Tapirs

The tapirs are recognized by the short moveable, prehensile snout, which forms a small trunk. There are 4 species, 1 dwelling in the forests of Malaya and Sumatra in Asia, and the other 3 in Central or South America. They are the only odd-toed ungulates which have an even number of toes! The front feet sport 4 toes, this being a primitive trait inherited from the

Malayan tapir

group's ancestors. The hind feet have 3 toes as do both the hind and forefeet of the rhinos. These heavy-bodied mammals have short, stout limbs and a short tail. Hearing and smell are well developed but the eyesight is poor.

Above: *The white rhinoceros of Africa is adapted to grazing grasses, having a very wide flat-lipped mouth.*

The Malayan tapir (*Tapirus indicus*) is the most striking in colour. The head, neck, shoulders and legs are black but the back, sides and belly form a dramatic white band. In the twilight or moonlight in its forest home, this no doubt helps to disrupt the tapir's shape and so aids in concealment. Within its range it travels along well established routes and marks them by depositing its droppings at regular intervals. Rhinos also do this: it is the standard method of marking territory for both these animals.

The South American tapir of the Amazonian forests is brownish-black in colour and has a short stiff mane. It is an excellent swimmer and will dive into the nearest available water if danger, such as a jaguar, threatens. The other two American tapirs prefer the higher slopes of hills and mountains.

After a long period of gestation, of about 13 months, tapir mothers usually give birth to a single youngster, but sometimes twins are born. The juvenile has longitudinal stripes and spots on the body and legs, which camouflage it amongst the ground foliage and dappled jungle light. These markings disappear when the animal is half-grown.

Horse family

The horse group forms the largest of the three families of odd-toed ungulates and is made up of the zebras, asses, and true horses. These animals all have larger eyes and thus better sight than rhinos and tapirs, but hearing and smell are still of greater importance. Most species of this group live on plains or in open environments, usually in herds that have a leader. As soon as danger is recognized their rapidly moving, long legs, powerful hearts and spacious lungs enable them to make a quick get-away.

Zebras are found only in Africa, south of the Sahara desert. They are easily recognizable with their black or dark-brown stripes on a whitish coat, but it is more difficult to differentiate between the 3 species. The mountain zebra (*Equus zebra*) is the smallest and has the most southerly distribution. Although it once roamed in huge herds, it is now restricted to protected areas in Cape Province. It is the most donkey-like of the zebras with long ears, narrow hooves and a heavy head. The stripes are narrow and close-set, except on the hind-quarters where they are broader and more widely spaced. As its name suggests, it is most agile on rocky terrain.

The common zebra (*E. burchelli*) is the most numerous zebra in Africa, and several races are identified from different areas. The markings are very variable but usually there are shadow stripes between the distinct zebra stripes.

Grevy's zebra (*E. grevyi*) is the largest of the zebras and has the narrowest stripes. It gives ass-like brays as distinct from the common zebra's barking call.

In their way of life the various species of zebra differ little from one another. The herd is usually led by a wise old stallion who is always on the look-out for any of their numerous predators, such as lion, leopard, hyena or hunting dog. He is especially alert when leading his herd to water at dawn or dusk. All zebras are sociable grazers and are often found mixing with other animals such as gnu, eland, hartebeeste and ostrich. The animals co-operate in giving warning of enemies.

A pregnant female leaves the herd briefly to give birth to a single, well developed baby. The youngster soon struggles to its feet, takes a nourishing drink from its mother's teats, and then follows mother to rejoin the herd. Joining the herd as soon as possible gives greater protection to the newborn, which are otherwise easy prey for the African carnivores.

The quagga (*E. quagga*) of South Africa was a zebra that was striped only on the head, neck

Right: Stallion zebras battle on the plains. In males the canine teeth are pointed and can inflict hard pinches in fights that are usually for the favours of a mare. Hind feet will be used to kick the opponent.

The domestic horse

The domestic horse has been living in association with man for more than 4,000 years. It is uncertain whether or not Przewalski's horse is the type from which our domesticated horses have been bred. There were probably several ancestors, at least three accounting for our modern breeds of horses, for three types stand out from today's numerous breeds, and they probably indicate the nature of the original ancestors. The first type includes breeds such as the Celtic pony, a small race of horse, and the Shetland, Icelandic and Norwegian ponies. The second type is made up of the 'hot-blooded' Oriental races, the long-legged breeds including the Arab, Hanoverian, Fullblood, and the English Thoroughbred. The English Thoroughbred or racehorse is of almost pure eastern blood. Modern thoroughbreds show a great deal of variation in type chiefly because there has been little or no selection except that based on racecourse performance. The heavier, western, heavy-legged breeds make up the third type and are mainly used for heavy farm work, such as pulling agricultural machines or carts.

The beautiful Shires, Belgians and Jutland horses are examples of these powerful and fine breeds.

Above: *A thoroughbred mare and her foal. These horses are bred for racing and can move at 16 metres per second.*

and shoulders. It became extinct in the 1870s, as a result of having been ruthlessly shot because it competed with domestic animals on the open plains of South Africa.

Asses, donkeys and wild horses

The asses have long ears, a short, erect mane running from between the ears to the shoulders, a thin tufted tail and long slender legs. There are 2 species, the Asiatic onager, kulan, or wild ass (*E. hemionus*) and the African wild ass (*E. asinus*). The African species is the ancestor of the domestic donkey or ass. Horses and asses will interbreed, the cross between a male ass and a mare being known as a mule, and that between a stallion and a female ass being known as a hinny. A zebroid is the offspring produced by an ass and a zebra. All these hybrid offspring are sterile.

There is only one wild species of horse surviving in the world today. This is Przewalski's or the Mongolian wild horse (*E. cab-*

allus), which survives in the cold inhospitable plains of Mongolia in central Asia. It differs from the breeds of domestic horses in having a short, stiff, erect, blackish mane, a heavy head, smallish ears and a low-slung tail. It has a most attractive pale reddish brown coat with darker legs and a white muzzle. In winter the coat grows long and shaggy to keep the animal warm on the cold hostile plains.

Below: *Przewalski's horses are the only wild horse species surviving in the world today. They inhabit the vast plains on either side of the Altai mountains in Mongolia.*

Even-toed ungulates

(ORDER ARTIODACTYLA)

The even-toed ungulates or artiodactyls are much more successful
than their odd-toed relatives.
There are 194 species distributed
almost world-wide, only Australia, New Zealand and the polar
regions being outside their range.

The identifying feature of this
group is that the central axis of
the foot passes between the third
and fourth digits of the toes. This
means that the pressure of the

body is taken in each foot by 2 or 4
hooves. The even-toed ungulates
have held their own in many parts
of the world, despite being hunted
by man.

There are two main groups, the
non-ruminants and the ruminants. The non-ruminants are
those that do not chew the cud
and have a fairly simple digestive
system with a two-chambered
stomach, and they include the
pigs and hippopotamuses. The
more advanced ruminants chew
the cud and they have a complicated four-chambered stomach.
This group includes deer, giraffes,

Above: *Wild boars of Eurasia snuffle with
sensitive snouts for fallen nuts and other
edible food. The young are striped,
possibly for camouflage purposes or for
family contact.*

antelopes, cattle, sheep and
camels.

In the ruminants, the food is
torn up and swallowed quickly. It
passes into the first two chambers
of the stomach, known as the
rumen and reticulum. Later,
when the animals are resting
safely away from predators, this
food is brought back into the
mouth in small lumps. Here it is
thoroughly chewed and then re-

swallowed. This time the food passes into the third and fourth chambers of the stomach, which are known as the omasum and abomasum. Eventually all the food is completely digested. The advantage of this system is that it allows the ruminant to obtain as much food as possible in a short time, and then retreat to a safe place to consume it at leisure. This is probably the main reason why the group is so successful.

Even-toed hoofed mammals often have paired horny outgrowths from the top of the skull. They are different in shape and structure, depending on the group and species.

Pigs, hogs and peccaries

The 8 species of wild pig do not look much like the domestic pig. The wild species are very hairy and often the upper and lower canines grow very long, forming sharp, curved tusks. Most domestic forms are rather scantily haired and do not bear tusks. Pigs are unusual among the ungulates in that the females or sows give birth to large litters and consequently have several pairs of mammae to provide milk for the youngsters.

The wild boar (*Sus scrofa*), was once plentiful in the woodlands of Eurasia, North Africa, Sumatra and Java. It has been intensely hunted through the centuries. Today the largest populations survive in various parts of Asia. These omnivorous ruminants move about in small parties called 'sounders'. The piglets are most attractive with horizontal stripes running along their brown bodies. It is thought by most zoologists that this colouration assists in concealment of the youngsters. However, recently it has been suggested that the marks are important visual signals enabling the mother to keep sight of the youngsters. This is born out by the fact that when she screams, as danger threatens, the young all run to her rather than crouch motionless.

The warthog (*Phacochoerus aethiopicus*) is the clown of the African open woodlands and acacia-spotted savannas. The male is rather ugly with a warty face and large tusks. Its sparsely haired body is supported on short legs. A small family party running along provides an amusing sight: the long thin tails, with a tuft of hairs on the end, are held erect, looking rather like bedraggled flags. The female has only 4 mammae so that litters usually number only 3 or 4.

The largest living wild pig is the giant forest hog of the rainforests of Kenya and the Congo. The strangest member of the pig family is the almost bald, wrinkled-skinned babirusa of the Celebes. Its native name means 'pig-deer' and refers to the huge antler-like tusks of the male, both upper and lower canines growing to great length.

Peccaries or javelines are found in Central and South America. Although they are superficially like the pigs of the Old World, they differ internally in a number of important details, and are consequently placed in a separate family. Differences include the fact that the upper tusks curve downwards (up in pigs), the hind limbs have 3 toes (4 in pigs) and the stomach is more complex.

Peccaries live in small family groups or huge herds of 100 or more individuals. They inhabit mainly forest edges, feeding on various plants and their fruits. Occasionally they will eat small animals such as rodents, insects, worms and even snakes.

The collared peccary (*Tayassu tajacu*) lives as far north as the deserts of the south-western United States. It acquired its name from the whitish broad stripe that extends round its throat like a loose collar. The white-lipped peccary (*T. albirostris*) ranges from Paraguay to Mexico, and has a white marking running from its chin nearly to its eye.

Below: *A disturbed African warthog runs from its mudbath with its thin tail held erect in typical stance.*

Hippopotamuses – the river horses

Hippopotamuses are distant relatives of the pigs. The hippopotamus (*Hippopotamus amphibius*) rivals the great Indian rhinoceros, as the second largest living land mammal. A mature male can be up to 4.3 metres long, some 1.5 metres high at the shoulder, and can weigh up to 4 tonnes. The enormous, thick-skinned,

Left: *Hippopotamuses resting on an African sandbar. Although rather clumsy on land where they graze nightly on vegetation, these distant relatives of the pigs are quite agile and graceful in water.*

River-horse herds

Its Latin name means literally 'river-horse' and during the sunny African days the hippo spends most of the time dozing in the water or basking on a sandbar. In the water, it can submerge for up to 4.5 minutes, but usually it just leaves its ears, eyes and nostrils showing above the water. Sometimes its back creates a stepping stone or landing point for river birds. At night, however, these heavy mammals come ashore and they look rather clumsy as they move off to graze on tall grasses and reeds. Hippos establish regular tracks over the ground, and usually they do not wander more than 2 or 3 kilometres from water but occasionally they may wander for anything up to 30 kilometres. Just before dawn they return to their river haunts.

Huge herds of 100 or more animals used to be a common sight, but today 'schools' as they are called, are greatly reduced. It was long thought that a school was led by a strong, dominant male, but studies have shown that the leader is in fact a matriarch. Each group establishes an aquatic territory, the central crèche being occupied by females and juveniles. Refuge areas around the edge are each occupied by an adult male. In the breeding season males fight for these areas and the fierce yawning display, often seen in zoo hippos, is actually territorial defence. Sometimes heated battles take place with the curved tusks being used to inflict gashes and wounds on the opponent. These quickly heal but scars remain. A female in season will choose her partner and enter his refuge to mate. About 240 days later the infant

hippo is born, almost 1 metre long and about 28 kilograms in weight. Although sometimes the infant is born in the water, usually the female prepares a trampled bed of reeds for the birth. Within minutes, the new-born hippo can stand and soon it enters the water, where it is able to swim well. It is taught to swim level with the mother's shoulders so that it is in a

protected position if danger threatens. On land it walks beside her neck so that she can keep a watchful eye on it.

The pygmy hippopotamus does not live in large herds, being found instead either singly or in pairs. This species spends more time on land, sleeping during the day and wandering the forests in search of tender shoots, leaves and fallen fruit, at night.

Left: *Arabian camels, the one-humped species, being herded across desert tracks, on their way to market. This species and most two-humped camels no longer exist in the wild.*
Below, left: *Camel portrait. A double row of interlocking eyelashes protects the eyes from too much glaring sun, while the slit-like nostrils can be closed at will to keep out dust or sand. Heavy eyebrows also help to protect the eyes in severe sandstorms that occur quite frequently in desert areas.*

almost hairless body is supported on short, pillar-like legs, each with 4 toes which end in hoof-like nails. The only parts of the body that are hairy are the inside of the ears, the muzzle and the tip of the tail. The tusks of the male grow continuously and usually curve, reaching between 40 and 80 centimetres in length. Once numerous in rivers throughout Africa, the hippo is now extinct north of Khartoum and south of the Zambesi river, except in protected areas such as the Kruger National Park. The other species of hippo alive today is the pygmy hippopotamus (*Choeropsis liberiensis*) which lives in the forest streams of West Africa. It is much smaller, the body being more rounded in shape than that of the common hippo.

Camels – the ships of the desert

There are 2 species of camel alive today, the bactrian and the Arabian camels. The two-humped or bactrian camel (*Camelus bactrianus*) inhabits the rocky deserts of central Asia, such as the Gobi Desert of Mongolia. In winter these deserts become snowy wastes. Although many two-humped camels are to be seen in zoological collections, most of these are of domesticated stock. Wild ones, of which there are very few, have smaller humps, smaller feet, shorter hair and are more slender in build than the typical domestic form, which was domesticated by the year 857 BC.

The Arabian camel (*C. dromedarius*) is easily distinguished from the bactrian by its single hump. It is unknown in the wild state but domestic forms are found living wild in Africa, Australia and southern Asia. Its original home was Arabia and the Sahara Desert borders. The dromedary is a special breed of Arabian camel suited for racing and riding.

Camels are well adapted for both cold and hot desert conditions. They are noted for being able to survive for very long periods without water. Fat, not water as commonly believed, is stored in the hump and this can be broken down within the body to give energy and water when required. The feet have 2 toes and undivided soles that prevent the animal sinking into soft sand in the case of the Arabian camel, or into deep snow in the case of the bactrian camel. The nostrils can be completely closed to keep out sand, and a double row of interlocking eyelashes protects the eyes against sun, sand and glare. The skin has no sweat glands, and this helps to prevent loss of moisture from the body.

Camels have been used for centuries as beasts of burden. They also provide hair for clothing, hides for leather, and milk and flesh for food. Even their dried droppings can be used as fuel for fires. They can still be seen travelling, laden with goods, in great caravans in many parts of the domestic range. Loads of up to 450 kilograms can be carried by the stronger individuals.

South American camel relatives

The 4 camel-like animals found only in South America are the llama (*Lama glama*), alpaca (*L. pacos*), guanaco (*L. guanacos*), and the small vicuna (*L. vicugna*). They do not have humps and are more slender and lighter in build than their Old World relatives. The llama and the alpaca have been domesticated since the time when the Incas lived in Peru. Today they no longer survive in the wild but the Peruvian mountain Indians keep them in large herds for their fur, hide, flesh and droppings. The llama is used mainly as a pack animal, while the alpaca provides the bulk of the wool. The wool is warm and light and very useful for rugged outdoor clothing. In recent years Peruvian clothing has become very fashionable in America and Europe.

The guanaco is the tallest South American mammal surviving in the wild today. With its long neck, it can reach up to about 109 centimetres in height. It is found at all altitudes up to 4,000 metres, although it is most common in semi-desert and high altitude plains. Small herds of 4 to 10 females and young are led by a mature male. Young males, on maturing, are chased away, and they join together in all-male herds of 12 to 50 individuals. During the mating season the leader has not only to keep his females together and mate with them, but he has also to continually chase away males who are without a harem and who try to steal his females. The female gives birth after 11 months gestation to a single young, which is able to stand and run with remarkable endurance very soon after birth.

Mouse-deer

The mouse-deer or chevrotains (*Tragulus* species) are some of the smallest of all living hoofed animals, standing about 30 centimetres high. They inhabit tropical forests and mangrove thickets. There are 4 species to be found – the water chevrotain of Africa, the lesser and larger Malay species, and the spotted Indian chevrotain from southern Asia.

Chevrotains are shy, wary and solitary, rapidly darting through the brush like rabbits. They mainly eat fallen fruit and leaves, often those of water plants. These animals are perhaps a link between the ruminants with a four-chambered stomach and the artiodactyls with a non-ruminating stomach, since they have a three-chambered stomach. An interesting detail in the male is provided by the extremely long upper

Below: The graceful guanaco, South American wild relative of the camels, lives in the high mountains where it feeds on the grasses of the mountain slopes. It enjoys standing and even lying in water.

Saved from extinction – the vicuna

The vicuna is a remarkable success story. It looks like a smaller, more slender version of the guanaco and it is found along the Andes on the high plateau between 4,250 and 5,500 metres. It is protected against the harsh winds, hail, frost and snow by its thick coat of fine wool. Until the middle of the twentieth century vicunas were wantonly slaughtered for their skins and wool, and it was therefore very surprising that in 1950 there were still some 100,000 animals remaining from the original population of one million or more animals. Despite conservationists' warnings overhunting went on until, in the early 1960s, their numbers had plunged to below 10,000. Peru fortunately established the first vicuna reserve and later Bolivia, Argentina and Chile helped to improve the vicuna's prospects. There are three action programmes — protection in the field, educating the school children and general public, and the banning of trade in vicuna products. Britain and the United States have banned the import of vicuna products and thus helped towards its survival. Numbers have increased to at least double the 10,000 figure. In some reserves the populations have increased to such large numbers that some animals can now be reintroduced into those places where their ancestors had been slaughtered long ago. Also the population in some reserves is naturally spreading outwards.

canines that protrude from the upper lips to form tusks. Neither sex possesses either horns or antlers.

Although quite numerous and well-known to natives, these shy, retiring, nocturnal beasts are seldom seen. As a result little is known of their behaviour in the wild, but the African species is reported to climb up a sloping tree to sunbathe or to escape predators. The gestation period is about 150 days after which a single offspring or occasionally twins are born.

Right: *The dimunitive mouse deer or chevrotain of Asia have legs about the size of an ordinary lead pencil. It is more closely related to the hog family and camel family than the deer family.*

Musk deer and muntjac

The single species of musk deer (*Moschus moschiferus*) is found in central and north-eastern Asia, from China and Korea to western Mongolia. It is a small, stockily built deer with a small head, large feet and very coarse hair. The back legs are longer than the front ones so that the back slopes up to the haunches. The head is rather reminiscent of a kangaroo's. Like the chevrotains, the males have long, upper, tusk-like canines. The males also bear a musk-secreting gland, which gives rise to their common name. The gland is found on the abdomen and secretes a brownish wax-like substance. It is this substance that provides one of the musks used in the manufacture of perfume and soap. Natives and hunters trap these deer in such large numbers that it is surprising that the species is not extinct.

The muntjac genus consists of about 6 species, that range south through India and Ceylon, east to Sumatra and its islands, and north to China and Formosa. An alternative name for these animals is barking deer, for when alarmed they utter a series of short, rapid barks which sound rather like castanets clicking. They are rather small, slender deer, being up to 1 metre long, standing about 0.5 metre high and weighing about 14 to 18 kilograms at maturity. The males have tusks like the chevrotains and musk-deer and also small, short, simple antlers. They prefer hilly ranges and areas of dense vegetation.

Eurasian deer

There are some 18 species of Eurasian deer, most of which frequent southern Asia and its various islands. These include the axis deer (*Axis axis*) and the sambar (*Cervus unicolor*), both of south Asia, and the sika deer (*C.*

Above: *A muntjac resting. This small, slender deer has short antlers carried on fairly long, bony, hair-covered pedicels.*

nippon) of south Asia, Japan and Formosa. The red deer (*C. elaphus*) and the fallow deer (*Dama dama*) are the only 2 species that are found naturally wild in Europe. Their range also extends well into Asia. Because they are most attractive and provide 'sport' for huntsmen, various deer species have been introduced into many European countries through the centuries.

The fallow deer is one such example. It originated from Mediterranean countries but today is very common in parks and forests all over Europe. A medium-sized deer, it stands just under 1 metre at the shoulder, and is most attractive. The spotted, fawn, summer coat moults to a uniform grey in winter. The broadly flattened antlers are a distinguishing feature. The male has a very prominent Adam's apple and the rutting cry is a very deep-

Right: *Male fallow deer fighting during the rutting season. During September and October, the mating season, the buck does a dance-like ritual and bellows in a deep voice, to attract females.*

toned grunt. One offspring is born in June or July.

One of the best known deer is the handsome red deer. An adult stag stands about 1.3 metres high at the shoulder and may have a full set of twelve-pointed antlers reaching another 1.2 metres in height. This is a 'royal stag'. Females are slightly smaller and do not grow antlers.

Stags live alone or in male herds during spring and summer but in autumn, with their antlers free of velvet and their coats in breeding bloom, they establish territories and gather their harems. This period is known as the rut. The stags emit deep powerful bellows to establish territories and attract a female. Fierce, aggressive battles between rival stags often take place, and the clash of antlers can be heard from a great distance. One spotted calf or fawn is born the following May or June after about 250 days gestation.

Père David's deer (*Elaphurus davidianus*) is now extinct in the wild, although it is still numerous in semi-captive herds. This is because the French missionary, Père Armand David, sent some of the Chinese emperor's herd back to Europe in 1865. The entire Chinese herd was destroyed during the revolution at the turn of the century. However, the late Duke of Bedford had established a very successful breeding herd on his estate at Woburn Abbey in England. Today these deer have been sent to many zoos and parks throughout the world.

Right: *Lord of the glen, a handsome red deer stag emits his deep, powerful bellow that echoes a long way. The sound defines his territory, keeps his females within his sight and warns other stags, tempted to enter his area, to be on guard.*

American deer

The wapiti (*Cervus canadensis*) of North America is more closely related to the Eurasian deer than the 12 or so other species of North and South American deer. It is about the same size as the red deer, but is greyish brown and the male has a dark chestnut mane. It browses and grazes on plants and twigs. It is territorial in behaviour and has a breeding pattern similar to that of the red deer.

The white-tailed deer (*Odocoileus virginianus*) and the mule deer (*O. hemionus*) are well-known American deer. The white-tailed deer is found from southern Canada south to northern South America, while the mule deer inhabits the western side of North America as far south as northern Mexico. They live in a variety of habitats but prefer areas with sufficient vegetation to give them cover. They do not congregate in herds, although 2 to 4 animals

Wapiti
(*Cervus canadensis*)

Moose and elk — different names, same animal

When animals of the same species are found in different countries it often happens that they are given different common names. This is the case with the species *Alces alces* of the timbered regions of northern America and Eurasia. It is known as the moose in America, and the elk in Europe. A stately mammal, it is the largest member of the deer family, and is easily identified by its broad, overhanging muzzle, massive spoon-shaped antlers, heavy mane, and the characteristic hanging flap of skin beneath the throat which is known as the 'bell'.

Moose live in small groups in summer, but when food gets scarce as the winter snows fall, they congregate in 'yards'. These are special areas in which they have trampled all the snow. In deep snow they are sometimes overcome by pumas and wolves, although in better conditions their antlers and hooves provide excellent defence.

may be found together. Buckskin, the leather tanned by the North American Indians, was made from their hides.

South America is the home for many little-known deer, including swamp deer, pampas deer and Andean deer. They live mainly in the areas suggested by their common names.

The semi-domesticated migrating reindeer

The reindeer of Eurasia and the caribou of North America are the same species (*Rangifer tarandus*). They are distinguished from all other species of deer by the fact that both the males and the females grow antlers. They are native to the arctic regions of the world and have been known to man since Palaeolithic times. The peoples of Siberia and the Laplands have semi-domesticated them for centuries. However, their urge to migrate with the changing season remains as strong as ever and this has caused the Lapps and native Siberians to share a nomadic life with the reindeer.

Throughout their range the reindeer gather in herds of hun-

Above: *A reindeer herd. This species is native to the Arctic regions of the world and is called caribou in America and Siberia.*

dreds or thousands for their annual migrations and move south to more favourable lands. Huge herds of 200,000 animals have been recorded trekking between summer and winter quarters. The rutting season occurs during the autumn migration, the fine branched antlers being shed in November. The fawns are born usually at the end of the return trip, the following late spring and early summer.

The okapi — the pigmies' donkey of the jungle

For a long time it was thought that the giraffe was the sole representative of a unique family. But in 1901 Sir Harry Johnston, Governor of Uganda, discovered a new mammal in the Semeliki Forest in the Belgian Congo. He remarked 'It has been one of the surprises of the twentieth century that a mammal so large and so eccentrically coloured could have remained unknown to science'. Sir Harry obtained only two skulls and a skin of this new mammal, but he did establish that the okapi (*Okapia johnstoni*) really did exist in the gloomy rainforests of central Africa. Outwardly the giraffe and the okapi are not very similar. The okapi is more like a horse than a giraffe and the name 'okapi' is a Pygmy one for 'donkey' or 'ass'. However, the internal anatomy does indicate the close relationship between the two mammals. The okapi is reddish brown in colour with white horizontal stripes on the rump and legs. The male has a pair of small 'horns'.

Not much is known of the okapi's habits except that it usually lives alone or as one of a pair, evading most danger with its keen sense of smell and hearing. It was thought to be nocturnal but this misconception arose because of its secretive, skulking habits.

Giraffe

The giraffe (*Giraffa camelopardalis*) of African savannas is the tallest of all living animals, standing at about 6 metres. It has overcome all competition from other African browsers by an immense elongation of its neck. Surprisingly the number of vertebrae in the neck remain the same as in most mammals, including man – a total of 7. On the top of the head are 2 to 4, short, blunt horns or knobs. Occasionally there is a 5th bump between the eyes.

The relatively long neck causes a problem when the animal takes a drink of water. To reach the ground it usually spans the front legs very wide, or alternatively it has to crook its knees. In this position it is very vulnerable to

Left: *Giraffe crèche. These 4 young giraffes each have their own mother but are left with an 'aunt' in a nursery while the adults go off to feed. The long legs enable them to move at speeds greater than those of the swiftest horses.*

attack by lions. It is found south of the Sahara, mainly in areas where acacia trees abound, since its main food intake is the leaves and horny twigs of this tree. Surprisingly the mobile lips and the long, coiling tongue seem immune to the acacia's long thorns.

If you have seen giraffes in zoos or parks, or been fortunate enough to visit the various game reserves throughout Africa, you no doubt will have observed different colour patterns on the skin of the giraffes. This is because several subspecies or races have evolved in the various regions. The reticulated race from East Africa has triangular patterns whereas the South African Cape giraffe is much more blotchily marked.

Giraffe family life

Despite their size, giraffes are shy, timid animals living quiet and inoffensive lives. They are gregarious, wandering about in herds of about 12 to 15 animals, although as many as 70 are sometimes seen in one herd. A small herd will usually include a mature male leader, cows and their calves of various ages, and some adolescent males not yet sufficiently mature to compete with the dominant head bull. The bull will keep the youths in order, as well as any intruding male, by using his long neck and head as a club. During courtship a male and female will rub their necks together and swing them from side-to-side, an act often called 'necking'.

The baby giraffe is born after a long gestation period of some 420 to 450 days. It weighs about 60 kilograms and stands from 1.7 to 2 metres high. The young giraffe is able to stand on its wobbly legs about 20 minutes after birth and it finds its mother's milk within about an hour. As they get older, the young giraffes gather in groups and are looked after by 'aunties' in a crèche. The mother feeds her young until it is about 9 months old.

The pronghorn – the fastest mammal in North America

The pronghorn (*Antilocapra americana*) is an antelope-like creature of North America. Its branched horns are shed annually. It can achieve speeds almost equal to those of a cheetah over short distances, that is over 90 kilometres per hour. Over longer distances it can maintain about half this speed for several kilometres. This would leave even a cheetah gasping for breath and several kilometres behind. To help it move, the pronghorn's hooves are covered with cartilaginous pads.

The pronghorn lives in small herds in wild, rocky desert country where the creamy colour blends in with the background. Coyotes are its main enemy, particularly in snow when the pronghorn cannot move quickly. It is hunted by man in some areas. Bobcats may also take the young.

Above: *Not true antelopes, American pronghorns grow horns in both sexes. They are the swiftest animals of the New World, cruising at about 50 kilometres per hour, and 90 over a short stretch.*

Bovids – cattle and their relatives

All the remaining hoofed mammals with an even number of toes have been lumped together into one vast family named the Bovidae. There are about 115 species in the family, among

Domestic cattle

Among the creatures most useful to man are the various breeds of domestic cattle (*Bos taurus*). They have served man since prehistoric days as beasts of burden, powerful draught animals and as suppliers of milk, meat and leather. The origins of domestic cattle are uncertain but it seems likely that the long-haired wild auroch is one of the ancestors of modern cattle. Julius Caesar refers to it in his writings and it was found throughout the forests of Eurasia and North Africa until it became extinct in the seventeenth century.

Domestic cattle have been developed into many breeds for production of milk, beef or both. Some of the breeds are illustrated here, but they are only a few representatives of the many established breeds found in the world today.

The Normandy is a valuable beef and dairy bovine. It is generally accepted that a Normandy cow can yield 27,500 litres of milk a year, from which can be made well over 150 kilograms of butter. The Jersey is a most attractive dairy breed with a beautiful honey coat and large appealing eyes. The aim with this breed is for the milk to be as rich as possible in cream. The somewhat larger Guernsey breed also produces rich, creamy milk.

A very common beef breed is the Shorthorn and it is one of the most widespread breeds in the world. The Aberdeen Angus yields some of the best quality beef. The wide-horned, shaggy Highland cattle are beef animals, but today they serve mainly as a tourist attraction in the Highlands of Scotland.

Another species of domestic cattle is the zebu (*Bos indicus*), called the brahman in Australia and the United States. This species flourishes in humid or arid districts and shows good resistance to infectious diseases.

which are such well-known creatures as cows, sheep, goats, antelopes, and gazelles. They are found throughout the world today, having been introduced into certain areas such as Australia and New Zealand.

True oxen

The ox is typical of the bovid family and especially so of the cattle group. It has a massive, heavy body with a short neck from which hangs a dewlap. The broad, bare muzzle is moist and the animal has an excellent sense of smell. The horns are either smooth or grooved just at the base. The wild species of ox living today all belong to the genus *Bos* and include the gayal, gaur, and banteng, which are all Asiatic and differ from the European domestic ox (*Bos taurus*) mainly by having slight humps on their backs. They all have similar habits. They live in small herds of between 5 and 20 individuals, led by a mature bull. They rest early in the morning and mid-afternoon when they peacefully chew the cud, having grazed during the rest of the day. They normally sleep at night unless they are disturbed by man, when they become more nocturnal in habit.

The yak (*Bos grunniens*) is an unusual kind of ox that lives in the uplands of Tibet, in small numbers. It has been domesti-cated for centuries in Tibet. The domestic yak is smaller than the wild yak, and is white, reddish brown or black in colour, while the wild yak is blackish brown with white on the muzzle. The yak's identifying feature is the very long hair, on the flanks, that grows down to its ankles, forming a fringe.

Below: The yak or grunting ox lives in desolate areas up to 6000 metres in the colder parts of Asia. In Tibet it has been domesticated for centuries and it is used as a docile beast of burden as well as a source of meat and milk.

CATTLE BREEDS

Friesian

Brahman

Herefordshire bull

Texas Longhorn

Highland

Buffaloes

The true buffaloes are found in the Old World, but the word 'buffalo' has been used in the United States for the North American bison. The true buffalo of Africa (*Syncerus caffer*) was once very numerous, but outbreaks of rinderpest swept through the continent at the end of the last century and the population suffered badly. Today the buffalo is confined to reserves and the thick bush along the eastern coast of Africa. Although heavily built and well equipped with massive curved horns, it is still attacked by lions. However, it is usually only cows and calves that stray from the herd that are singled out. The African buffalo is sometimes called the Cape buffalo to distinguish it from the smaller and rarer dwarf forest buffalo, which inhabits Central and West Africa.

Bison

There are two species of bison, the North American bison (*Bison bison*) and the European bison or wisent (*B. bonasus*). The American bison was formerly a most abundant hoofed mammal ranging the prairies of North America in millions. As man opened up the West with railroads and towns the vast herds were exterminated. Today these striking animals can be found only in parks and reserves. They are huge beasts, with a massive, low slung head and large hump. The dark brown shaggy winter coat falls off in patches in spring and is replaced by a lighter, shorter one. The largest males reach a height of 1.8 metres at the withers and a weight of 1.5 tonnes. Bison graze, chew the cud, sleep and go to water-holes to drink, showing an outwardly placid temperament for

Left: *Water buffaloes* (Bubalus bubalis) *roam wild in north-eastern India, but the majority are domesticated animals.*

Massive migrations of the past

The American buffalo is famous for its annual migrations, when vast herds moved south to avoid the winter snows. They did not move as far as many writers have suggested. For example, it was said that herds found in Saskatchewan travelled as far south as Texas, but it is now thought that all the herds moved southwards just 300 kilometres or so to more favourable winter pastures. With a population of 40 to 60 million animals before the advent of Europeans this migration must have been an impressive sight. By 1800 the eastern populations were exterminated and by 1875 the western herds were reduced to isolated pockets. In 1893 there were only about 20 wild bison in the Yellowstone National Park and about 300 in the Wood Buffalo Park area of the Mackenzie district. However, numbers have considerably increased and their survival is ensured. The bison of Wood Buffalo Park, for example, still undertake considerable migrations. They move from the wooded hills to the Peace Valley River area, a distance of as much as 240 kilometres. They travel just over 3 kilometres a day on their journey.

Above: *American bison formerly occupied most of North America. Today, due to massive extermination in the 19th century, they survive in National Parks because of the efforts of a few public-spirited persons who saved the 500 or so surviving bison at the turn of the century. In future they may be farmed for meat.*

most of the year. In the autumn mating season, however, the bulls fight for possession of the cows. There is preliminary stamping, then they charge at each other, head on, coming together with a resounding crack. They will also charge and fight wolves that attempt to prey on young calves or cows.

The European bison is extinct in the wild but semi-wild herds have been established in the Bialowieza Forest in Poland, from captive specimens obtained from zoos and private collections. The European species is more graceful than the American one, with longer legs, smaller head, more slender horns and a less shaggy covering. It is a woodland species browsing on ferns, leaves and bark.

Right: *The European bison are also known as wisents. Their numbers were reduced to a few hundred in the 18th century.*

Left: *Typical defence formation put up by Arctic musk oxen against intruders such as man or an attacking pack of hungry wolves. Young are protected in the centre.*

Arctic musk ox

An impressive sight of the barren wastes of the Arctic is the musk ox (*Ovibos moschatus*). This long-haired cow-like animal is adapted to survive in the inhospitable tundra regions of Canada, Alaska and Greenland. It looks rather like a yak with the long shaggy coat that helps insulate it, but the head is very broad and low slung. These animals are usually found living in herds numbering between 10 and 30 animals, in winter. The herds split up into smaller groups in summer when food is not as difficult to find. They survive on mosses, lichens and dead grass in winter. In summer they enjoy fresh grass and the shoots of dwarf willows and scrub pines.

When it faces danger from a wolf, polar bear or hunting Eskimos, the herd forms a hollow circle, with the enormous expanded horns facing outwards. The young animals are protected in the centre. Although they look cattle-like, these bovids are more closely related to sheep.

Antelope and gazelles

Antelopes and gazelles are swift, graceful animals that live in Africa and southern Asia. There are over 60 species which are identified mainly by the shape of their horns. The size of antelopes ranges from that of a hare, the royal antelope (*Neotragus pyg-maeus*) for example, to that of an ox, the giant eland (*Taurotragus derbianus*) for example.

The different species are adapted for various habitats. Those that live on the plains, such as the wildebeest or gnu (*Connochaetes taurinus*), the eland (*Taurotragus oryx*), the springbok (*Antidorcas marsupialus*), and the impala (*Aepyceros melampus*) all form herds. Those that live in thick cover, such as the small duikers, royal antelopes and dik-diks, are generally solitary or else they live in pairs. The 3 species of oryx are adapted to living in the harsh deserts of Africa. They subsist almost entirely on seasonal grasses and the leaves and roots of stunted shrubs. They seldom are able to drink and obtain most of the moisture that they require from the roots and plants that they eat.

Most antelopes breed once a year. During the rutting season the males of herding antelopes become very territorial. Usually a single offspring is born. It can soon stand and run with its mother.

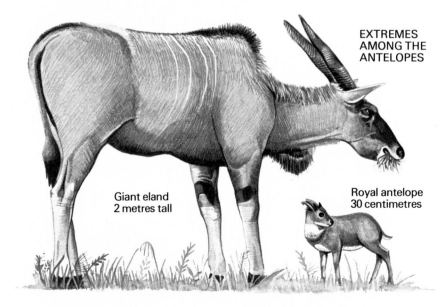

EXTREMES AMONG THE ANTELOPES

Giant eland
2 metres tall

Royal antelope
30 centimetres

Duikers – small African antelopes

The duikers form a large group of some 17 species of small antelope that are found in all parts of Africa. The most widely distributed of all the African antelopes is the grey duiker (*Sylvicapra grimmia*). The male and female are similarly coloured, the natural colour range being from fawn to greyish blue. Both usually bear straight, closely set horns, although those of the males are longer than those of the females, being up to 15 centimetres. Duikers have prominant facial glands. The scent from these is rubbed on to twigs and branches. In this way a duiker marks its home range by leaving its own personal smell. These antelopes browse on the vegetation, often standing on their hind legs to reach twigs and berries, but they also eat insects and snails. Forest duikers and blue duikers are close relatives.

Below: A duiker antelope at rest. The dark slit below each eye opens into a facial gland, the scent from which is smeared on to vegetation, thus marking its home range. When alarmed a duiker will dart at speed into dense vegetation and its name means diving buck. No duiker is more than 76 centimetres high.

Large antelopes

There are several large antelopes to be found in Africa. The giant eland is the largest, being heavily built and weighing up to 900 kilograms. Both sexes grow slightly spiral horns, bear a hump on the shoulders, and have a prominant dewlap. They are remarkably agile for their size and have been seen to jump over bushes some 2 metres high – about the same height as an adult eland. They live in small herds in South and East Africa.

The related greater kudu (*Taurotragus strepsiceros*) is also quite large, and very striking. The male's 'crowning glory' is a pair of long, corkscrew horns that can measure about 1.5 metres. The hornless female is half the size of the male. This antelope inhabits the rocky bushveld of Africa, moving about alone or in small parties consisting of cows and their young. Adult bulls are allowed to join the herd during the mating season.

A smaller close relative is the bushbuck (*T. scriptus*). This attractive antelope inhabits the dense bush and forest of tropical Africa. It is a rather secretive animal, preferring to live alone or as one of a pair in thickly wooded ravines along river banks.

The largest Indian antelope is the nilgai (*Boselaphus tragocamelus*). The common name means 'blue cow' but the female is in fact a brownish colour. Although it has been hunted for centuries, it is still fairly common in the northern states, living in small herds in open forest or parkland. Its only relative in India is the four-horned antelope.

Impala

The impala (*Aepyceros melampus*) is perhaps the most familiar and typical of all the antelopes. The male has lyre-shaped horns, stands almost 1 metre at the shoulder and weighs up to 37 kilograms. The female is hornless and slightly smaller. Both have reddish brown backs, with a lighter fawn band along the sides, and white underparts. This species is plentiful almost everywhere within its range on the savannas of South and East Africa. It browses and grazes, never wandering too far from water.

Right: Graceful impala hinds leaping over African thorn bushes. When running they can make successive leaps of 8 or 9 metres, often jumping over each other or springing into the air when there is no obstacle to clear.

The herds of impala range in size from a few to several hundred animals. A herd is led by an old male, with several young males acting as sentinels. When danger is sensed a sneezing alarm call is given. This sends the whole group bounding away in leaps that reach over 3 metres in height and cover distances of almost 9 metres.

During the rutting season the male struts around with his tail spread fanwise over his rump, displaying the vivid white patch underneath. A pregnant female gives birth, after about 171 days, to a single baby.

Below: The nilgai or bluebuck lives in forests and low jungles of peninsular India. Old bulls, such as the one seen here, prefer to lead a life by themselves for most of the year. The cows group with their calves in small herds. They browse and graze on the vegetation, but also take fruits and sugar cane, doing considerable damage. Hindus regard them as relatives to the sacred cow so they are not hunted.

The giraffe-antelope

The gerenuk (*Litocranius walleri*) is also named giraffe-antelope because of its long neck and its habit of making itself taller by standing on its hind legs. It does this to reach high foliage. In this position its usual, metre, height is extended to almost double. The long neck also gives the animal a better view of the predators, such as lion, leopard and jackal, that inhabit the scattered bush country of Somaliland and East Africa where it lives. When frightened the gerenuk moves off at a stealthy, crouched, trot with its neck held out horizontally, parallel to the ground. It is not as fast-moving as other antelopes.

Left: *The gerenuk demonstrates how it can reach quite high into an acacia bush. The acacia leaves are plucked with a long tongue and long upper lips in a manner similar to that used by the giraffe.*

Saiga success story

The peculiar bulbous nose of the saiga antelope (*Saiga tartarica*) easily distinguishes it from all other antelopes. This species roamed throughout Eurasia 20,000 years ago, but hunting reduced its range to the steppes of southern Russia, and at the beginning of the twentieth century it was in danger of becoming extinct. The reduction in the once numerous herds was largely due to the horns. These, when ground into powder, were considered a prized ingredient in the Chinese pharmaceutical trade. Fortunately, before it was too late the species was placed under strict protection, in 1920. The numbers have thankfully increased to such an extent that now individuals have to be culled each year to prevent over-population. There are well over one million animals living today.

Wildebeests

A rather curious and comical antelope is the gnu. There are 2 species, the brindled gnu or wildebeest (*Connochaetes taurinus*) and the almost extinct whitetailed gnu (*C. gnou*). The former is widespread over large areas of the grasslands of Africa and has an ox-like head and horns, bristly facial hair and a rather horse-like body. This makes it look rather ferocious and formidable.

On the Serengeti plains of East Africa the breeding and migration of gnu are keyed to the rains. During the rainy season these antelopes are widely spread, grazing on the green grass. As the dry season advances, the

Left: Migrating wildebeests. The animals mate while on this journey, and calve when they reach open savanna. Their greatest enemies are the lions.

herds congregate and move westwards to find new grazing areas. The animals mate during June while on this journey. The dry season is spent in wooded grasslands near the available water of rivers that have not dried up. As the rainy season approaches, the herds split up into smaller groups and return to the open savanna for calving. The young are born after about 8 to 9 months gestation, during the first rains.

Gazelles – graceful and agile

Two species of the graceful gazelle also form large herds. These are Thomson's gazelle (*Gazella thomsoni*) and the larger Grant's gazelle (*G. granti*). 'Thommies', as they are affectionately called, live right across the Serengeti, while Grant's gazelles inhabit the dry

Above: Springbok (Antidorcas marsupialis) herd. National emblem of the Republic of South Africa, it is native to the Kalahari Desert, Angola, and South Africa. At one time they migrated in millions but were almost wiped out.

Masai Steppe. Both are impala-shaped antelopes and both have a dark stripe down the side. The side stripe in Grant's gazelle is paler, however, and sometimes is absent altogether.

The springbuck or springbok of South Africa is a related species, similar to these gazelles, but it has a different tooth structure. The common name arose from its habit of leaping 3 to 3.5 metres into the air when startled or when at play. When 'pronking' the legs are held stiffly and close together, with the head lowered. When the antelope hits the ground it rebounds with effortless ease.

311

Domestic sheep

The origin of domestic sheep is rather uncertain. The sheep was, however, the next ruminant, after the goat, to be controlled by man. This was as early as 12,000 years ago in south-west Asia. It is thought the urial, the wild sheep of southern Asia, was the first sheep to be domesticated. However, the 450 breeds of domestic

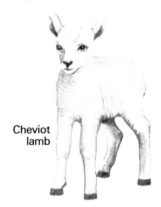

Cheviot lamb

Left: *Australian sheep shearing time. The Merino was originally bred in Spain and is famous for the fineness of its wool. It carries 5 times more wool fibres per square centimetre of its skin than any other related breed of sheep.*

sheep that are known today, probably come from several wild species that include the urial, the mouflon and the argali. The modern breeds have been selected for every kind of pasture, for most climates and for most altitudes.

The Merino gives the finest wool. It was originally bred in Spain, but today it is extensively farmed around the world, in those regions where a moderately dry, warm climate exists. New Zealand Mutton Lincoln breeds were crossed with the Merino to produce the all-purpose Corriedale sheep. The British South Down is another breed that is reared for its mutton and its fleece.

In central Asia, the Karakul sheep is raised to produce the commercial astrakhan fur. The three-day-old lambs are slaughtered for their black, tight-curled coats.

The Razza Sarda Sheep from Sardinia is one of the few breeds that are reared for milk production.

Lop-eared ram

Merino ram

Karakul ram

Hampshire ewe

Wild sheep

There are about 37 different species of wild sheep to be found in the world. The smallest are those of the mouflon group, and the largest of all living sheep is the argali which lives in the semi-desert regions bordering the Gobi Desert, in Mongolia. The only wild sheep to be found in Africa is the barbary sheep of the bare, rocky districts of the mountains of North Africa. The mouflon is the European wild sheep (*Ovis musimon*), living in Corsica and Sardinia. The bighorn sheep (*O. canadensis*) lives on the wildest and most inaccessible mountains

Left: *Bighorn sheep live in herds of about 50 in the mountains and have an unsurpassed ability for climbing and jumping. Only the rams have the big horns, the ewes' horns being much smaller.*

in North America and northeastern Siberia. The massive horns of the males curve into a full circle. When the rams are fighting during the breeding season they run at one another and crash the bases of their horns together. The sound echoes around the valley and it is amazing that they show no signs of injury – apparently not even having a headache.

Wild and domestic goats

The best-known goats are the wild goat or pasang (*Capra hircus*), of southern Asia and the various ibex races (*C. ibex*). The typical, Alpine, ibex once roamed the mountains of the Alps, but nearly became extinct last century. A law passed in 1920, making its territory a national park,

has saved it, and now once again it roams a wide area of the Alps. Several other races are found such as the Nubian, and the Siberian ibex.

Ibex are all short-legged, squat, sturdy animals with powerful horns, which are triangular in section and strongly rigid. The habits of ibex and other wild goats are very similar. They live on mountain crags and cliffs close to the snow line all the year round. They are extremely sure-footed and agile, and with their keen senses do not let predators or observers approach too close. They quickly move away and disappear over a ridge.

The domestic goat breeds probably originated from the Asian wild goat. They are extremely hardy, disease resistant, and easy to care for. In many underdeveloped countries they provide

313

Above: *The ibex is a wild species of goat, usually found living on rocky crags and mountain meadows of certain Eurasian mountains such as the Alps. It migrates to lower pastures during the winter when deep snows and severe weather hit the mountain tops.*

the main source of milk for the local people. Goat's milk is easier to digest than cow's milk and does not harbour tubercular infection. The domestic goat gives birth to 1 to 3 kids, twins being quite common, while the wild goat usually gives birth to 1 kid at a time, twins only occasionally being born.

The most important milk-producing breed is the Swiss or Alpine goat. The eastern or Nubian goat has large, drooping ears. The Kashmir and Angora goats are reared for their fine wool. The thick, white hair of the Tibetan kashmir goat consists of long guard hairs and a highly-valued silky underwool. In contrast, the angora goat from Turkey has long, silky guard hairs and it is these which form the valuable part of its coat. Today the mohair wool, which is spun from the fleece, is mainly produced in Texas.

Markhor goat
(Capra falconeri)
of the Himalayas
and Afghanistan

Goat-antelopes

Linking goats with antelopes are several species of quite rare mammal. They are called goat-antelopes because they look like goats, smell like them, and have their superb climbing abilities. Their faces and muzzles are usually more ox-like or antelope-like and they do not usually sport a beard. Some have horns that are like those of cattle, while others have horns that are like those of antelopes.

The only European species is the chamois (*Rupicapra rupicapra*), which lives in the major mountain peaks such as the Alps and Pyrenees, at heights of 800 to 2,800 metres. The herds are mixed and number between 10 and 50 animals which are very agile. They leap and run over and along rock faces that would defeat even

Above: *Chamois inhabit the mountain ranges in Europe and Asia Minor. When alarmed, these swift-footed and agile animals flee to the most inaccessible places.*
Right: *Goats were first domesticated between 8000 and 9000 years ago.*

the best mountaineer. When the winter snows arrive they usually descend and seek shelter in the forests.

In North America the Rocky Mountain goat (*Oxeamnos americanus*) is the only ungulate to keep a white coat all the year round. These animals live on the craggiest and most remote mountain slopes, usually well above the tree line. They live in small groups, the females retiring to quiet spots to bear their one or two offspring in April or May. Other species of goat-antelope are the gorals, serows and tahrs, which live mainly on the mountains of southern Asia.

Index

Page numbers in italic refer to captions to illustrations.